T0140761

On the Coordination of Saccades with Hand and Smooth Pursuit Eye Movements

Dissertation

zur Erlangung des Grades eines
Doktors der Naturwissenschaften

der Mathematisch-Naturwissenschaftlichen Fakultät
und der Medizinischen Fakultät
der Eberhard-Karls-Universität Tübingen

vorgelegt von

Hans-Joachim Bieg
aus Böblingen

Oktober 2013

Bibliografische Information der Deutschen Nationalbibliothek

Die Deutsche Nationalbibliothek verzeichnet diese Publikation in der Deutschen Nationalbibliografie; detaillierte bibliografische Daten sind im Internet über http://dnb.d-nb.de abrufbar.

©Copyright Logos Verlag Berlin GmbH 2014

Alle Rechte vorbehalten.

ISBN 978-3-8325-3648-0

Logos Verlag Berlin GmbH
Comeniushof, Gubener Str. 47,
10243 Berlin
Tel.: +49 (0)30 42 85 10 90
Fax: +49 (0)30 42 85 10 92
INTERNET: http://www.logos-verlag.de

Tag der mündlichen Prüfung: 7.2.2014

Dekan der Math.-Nat. Fakultät:	Prof. Dr. W. Rosenstiel
Dekan der Medizinischen Fakultät:	Prof. Dr. I. B. Autenrieth

1. Berichterstatter:	Prof. Dr. Heinrich H. Bülthoff
2. Berichterstatter:	Prof. Dr. Uwe J. Ilg

Prüfungskommission:	Prof. Dr. Heinrich H. Bülthoff
	Prof. Dr. Uwe J. Ilg
	Prof. Dr. Hanspeter A. Mallot
	Prof. Dr. Jean-Pierre Bresciani

I hereby declare that I have produced the work entitled: "On the Coordination of Saccades with Hand and Smooth Pursuit Eye Movements", submitted for the award of a doctorate, on my own (without external help), have used only the sources and aids indicated and have marked passages included from other works, whether verbatim or in content, as such. I swear upon oath that these statements are true and that I have not concealed anything. I am aware that making a false declaration under oath is punishable by a term of imprisonment of up to three years or by a fine.

H.-J. Bieg

ACKNOWLEDGEMENTS

I would like to thank Prof. Heinrich Bülthoff for his encouragement and support and Dr. Lewis Chuang and Prof. Jean-Pierre Bresciani for their guidance and friendship.

I would like to thank the members of my advisory board Prof. Heinrich Bülthoff, Prof. Uwe Ilg, and Prof. Hanspeter Mallot for their help and constructive criticism. I thank my colleagues, in particular Dr. ir. Frank Nieuwenhuizen and all members of CAPA, for an inspiring and delightful time.

I am also grateful to my parents for their support and encouragement. Most of all I would like to thank Sabine for her love and understanding.

ABSTRACT

Saccades are rapid eye movements that relocate the fovea, the retinal area with highest acuity, to fixate different points in the visual field in turn. Where and when the eyes shift needs to be tightly coordinated with our behavior. The current thesis investigates how this coordination is achieved.

Part I examines the coordination of eye and hand movements. Previous studies suggest that the neural processes that coordinate saccades and hand movements do so by adjusting the onset time and movement speed of saccades. I argue against this hypothesis by showing that the need to process task-relevant visual information at the saccade endpoint is sufficient to cause such adjustments. Rather than a mechanism to coordinate the eyes with the hands, changes in saccade onset time and speed may reflect the increased importance of vision at a saccade's target location.

Part II examines the coordination of smooth pursuit and saccadic eye movements. Smooth pursuit eye movements are slow eye movements that follow a moving object of interest. The eyes frequently alternate between smooth pursuit and saccadic eye movements, which suggests that their control processes are closely coupled. In support of this idea, smooth pursuit eye movements are shown to systematically influence the onset time of saccadic eye movements. This influence may rest on two different mechanisms: first, a bias in visual attention in the direction of pursuit for saccades that occur during smooth pursuit; second, a mechanism that inhibits the saccadic response in the case of saccades to a moving target. Evidence for the latter hypothesis is provided by the observation that both the probability of occurence and the latency of saccades to a moving target depend on the target's eccentricity and velocity.

CONTENTS

1

INTRODUCTION

1.1 WHY DO WE MOVE OUR EYES?

To see the world clearly, our eyes need to move. This may seem surprising since anyone experienced in photography knows that it is imperative to keep a camera still to take a sharp picture. Our eyes are not much different. As light travels through its optics it falls on the photoreceptors on the retina (see Fig. 1 A). As in a camera, it is important that the projection on the retina is stable to see the world sharply. Yet, unlike in a camera, the image is not processed uniformly across the retina. A tiny area of the retina delivers an image that exhibits much higher acuity than the rest (Fig. 1 B). This area is called the fovea and extends to an eccentricity of ca. 1 ° (Larson and Loschky, 2009). Thus, while it is also important to keep the retinal image still, it is equally important to shift it around, in order to foveate different areas of the visual field.

The eyes perform a range of movements to solve the conundrum that is posed by the need for stability and flexibility. Reflex-like eye movements stabilize the retinal image during body movements or other behaviors that induce large visual field changes, for example, the vestibular-ocular reflex or optokinetic reflex (reviewed by Ilg, 1997; Angelaki and Hess, 2005; Angelaki and Cullen, 2008). Another class of eye movements, *smooth pursuit*, stabilize the retinal image of objects that move through the visual field (Section 1.5). A class of eye movements that allow for flexible relocation of the fovea are *saccades* (Section 1.4). They rapidly move the eyes around to foveate different parts of the visual field in turn. A stable period lies in between two saccades wherein the eye remains relatively stationary and fixates the new location (but see Martinez-Conde et al., 2004). This intermittent behavior can be observed in many everyday activities, from reading (Rayner, 2009) to household work (Land and Hayhoe, 2001). Because of their high velocities, vision is impaired during saccades and only available afterwards, during fixation (Dodge, 1900; Zuber and Stark, 1966).

It is rather puzzling that we do not usually notice nor are we seemingly hindered in our daily activities by this piecemeal nature of vision. The brain must be very effective in determining where, when, and how a saccade should move the eyes in order to use their sparse resources in the best possible way. A part of the brain's ability to do so may stem from mechanisms that tightly coordinate saccade generation with other behavior that heavily rely on vision, for example, hand movements.

The current thesis presents a range of experiments that help us to investigate this issue. Specifically, work is discussed on the coordination of saccades and hand movements (Chapter 2) and on the coordination of saccades and smooth pursuit eye movements (Chapters 3 and 4). An overview and discussion of the experiments can be found in Section 1.6. After a brief motivation of oculomotor research from an applied perspective, the remainder of this introduction brings the reader in contact with the core concepts that are of importance in studying eye movements.

1.2 WHY STUDY EYE MOVEMENTS?

The fundamental role of eye movements for vision is certainly justification in itself to study how the brain controls these movements (von Helmholtz, 1867; Findlay and Gilchrist, 2003). In addition, understanding the principles of oculomotor behavior also provides for a range of innovative applications.

Simple bedside inspections of ocular behavior are employed routinely by physicians to examine common neurological damages (e.g., optic nerve damage, Yanoff and Duker, 2008). With modern video-based oculometers, more precise eye movement measurements can now be obtained relatively easily (Section 1.3). This, and a growing understanding of the processes in the brain that influence oculomotor behavior (Section 1.4 and 1.5) make it attractive to employ measurements of patients' gaze behavior for diagnostic purposes (Sweeney et al., 2004). This is not only useful for illnesses that are classically associated with motor deficits such as multiple sclerosis, Parkinson's

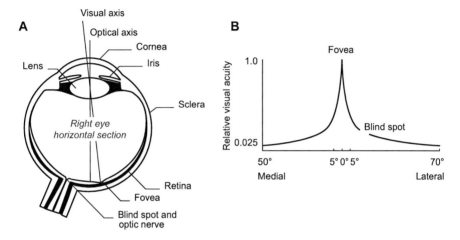

Figure 1: **A** Schematic of the eye, horizontal section of the right eye. Light travels through the main optic media of the eye, the cornea and lens, and is absorbed by photoreceptors on the retina. The fovea is a small retinal area close to the optic axis of the eye that has highest photoreceptor density and the lowest obstruction of photoreceptors by outgoing nerve fibers. Anatomical details from Cornsweet (1970). **B** Acuity measured at different locations in the visual field centered on the fovea of the right eye (adapted from Wertheim, 1894). This illustrates the acuity advantage of the fovea over other retinal locations.

disease, or myasthenia gravis (Serra et al., 2003; Leigh and Kennard, 2004; Tedeschi et al., 1991) but also for psychiatric disorders (e.g., schizophrenia), which are often multifaceted and difficult to diagnose (van Os and Kapur, 2009; Gottesman and Gould, 2003). In this respect, it is hoped that the targeted disease can be described by an array of clearly defined and measurable biological markers. Characteristic deficits in eye movement behavior, for example, increased interruption of smooth pursuit by saccadic eye movements in schizophrenic patients (Levy et al., 1993), could constitute one such marker (McDowell et al., 2011; Trillenberg et al., 2004).

The characteristics of unimpaired eye movements are at the focus of eye gaze measurements as they are used to study user behavior in ergonomics research and machine-interface design. Here, gaze data often complements other behavioral measurements (e.g., hand and body movements) to paint a finer picture of the user's interaction. A growing understanding of the connections between eye movements and these behaviors improves the analyst's ability to interpret and integrate these data in a meaningful way (Jacob and Kam, 2003; Goldberg and Wichansky, 2003). Furthermore, several research projects currently seek to employ on-line measurements of eye movements in an interactive way, for example, as a replacement for or in conjunction with traditional computer input devices such as the mouse or keyboard (Jacob, 1991; Salvucci and Anderson, 2000; Ward and MacKay, 2002). These interactive systems need to interpret eye movements in conjunction with other user actions to determine the correct user intention (Bieg, 2009).

These application areas can benefit from a growing understanding of how the eyes are controlled. Next, I will briefly consider the techniques that are required to study eye movements.

1.3 HOW TO STUDY EYE MOVEMENTS?

Eye movements have been studied throughout the history of mankind. An extensive review of early eye movement research is provided by Wade and Tatler (2005). The first mechanical devices that enabled a recording of eye movements date back to the turn of the 19th century. These devices required an actuation by the eye ball, which is, understandably, unpleasant for the studied observer and impractical.

Eye-tracking techniques today make use of either search coils, electrodes, or video cameras (reviewed by Duchowski, 2007). Video-based oculometry is the least obtrusive of the current techniques. Here, video cameras record one or both eyes. The rotation of the eye ball is then estimated using computerized image analysis techniques that automatically detect one or several features of the eyeball such as the pupil, limbus, or Purkinje reflections. Modern video oculometers sample the eye ball position with high frequencies (up to 2000 Hz, e.g., SR Research Eyelink 1000) and offer a data quality that is comparable to search coil recordings (van Der Geest and Frens, 2002). In fact, recordings of saccade kinematics have been shown to be more reliable using video-based techniques (Frens and van der Geest, 2002; Träisk et al., 2005).

Video-based eye movement recordings were also employed in the experiments presented in the current thesis using an SR Research Eyelink II infrared oculometer (sampling rate 500 Hz, see Fig. 2). In this system, video cameras are attached to a head band that is worn by the observer during the experiment. An estimate of the observer's eye position is derived from the pupil position in each camera image. A calibration procedure is required for calculating the point of regard of the observer on the computer screen. This procedure was regularly performed during an experiment. To avoid recording errors due to slippage of the system on the head, a chin rest was employed during all experiments. This set-up allows for accurate measurements up to 0.3 ° visual angle.

1.4 SACCADES

Saccades are very brief, high-speed eye movements that move the retinal image such that a new scene location is projected to the foveal area (Javal, 1879; Wade and Tatler, 2005). Each saccade is followed by a fixation during which the visual information is transduced and processed. In everyday behavior, a saccade is executed approximately 3–4 times per second (Findlay and Gilchrist, 2003). The following

Figure 2: Basic set-up used in the experiments. A head-mounted eye-tracker (SR Research Eyelink II) was used in conjunction with a chin rest. In interactive tasks, participants either used SR Research's response box (depicted here), the keyboard, or a custom-built potentiometer joystick, which was mounted under the desk within easy reach of participants.

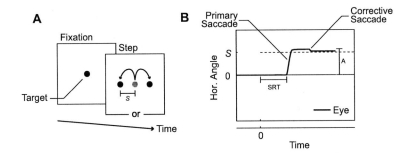

Figure 3: **A** Laboratory saccade task. The observer fixates on a central fixation object. After some time, typically, a random foreperiod between 1–2 s, the target undergoes a step and assumes its new position at eccentricity S. **B** Typical response to such a basic target step. The time before the observer initiates the saccade is the saccade reaction time SRT. Typical SRTs are in a rang of 200 ms. The size of the saccade in degrees is the saccade amplitude A. It does not necessarily correspond to the required amplitude. In the example, a small corrective saccade follows the main saccade to compensate for a slight overshoot.

section describes saccadic behavior in more detail and present several fundamental concepts that are of importance to the current thesis.

Where: Saccade Endpoint

Where the eye is moved to by the next saccade has been examined from very different perspectives. For example, it has been investigated how the brain *selects* which location in the visual field to fixate. Early research by Yarbus (1967) confirmed the intuitive assumption that this selection process is largely voluntary and determined by the goals of the observer. However, a second influence on saccade target selection is exerted by the distinctiveness or salience of the environment. For example sudden onsets of an object are very distinct and often lead to unintentional, reflexive eye movement (Theeuwes et al., 1998; Walker et al., 2000; Godijn and Theeuwes, 2003).

Where the saccade moves the eyes is also closely linked to the location of visual *attention*. We can attend to a location in the visual field without moving the eyes. Much like a spotlight (Posner, 1980; Posner and Dehaene, 1994; Brefczynski and DeYoe, 1999) the locus of this *covert attention* can be shifted such that different areas in the visual field receive processing enhancements, for example, higher contrast sensitivity or spatial resolution (Carrasco et al., 2000; Yeshurun and Carrasco, 1998). Investigations of attention in regard to saccadic eye movements show that covert visual attention and the location targeted by a saccade typically coincide (Rizzolatti et al., 1987; Hoffman and Subramaniam, 1995; Deubel and Schneider, 1996). Experimental dissociations, for example, by requiring observers to attend to one location but saccade to another, lead to impairments in both tasks (Kowler et al., 1995).

Independent from the target selection process, the second perspective on the *where* aspect of saccades concerns the actual *landing position* of the saccade. The distance that a saccade moves the eye is typically referred to as its amplitude. This is reported in visual degrees from the currently fixated location to the endpoint of the saccade, which ideally corresponds to the location of the saccade target (see Fig. 3 B). Frequently, both locations are mismatched and the saccade either overshoots (hypermetry) or undershoots (hypometry) the target location. Saccades tend to show hypometry to far away targets and hypermetry to targets that are close by (Carpenter, 1988; Kalesnykas and Hallett, 1994). Targets at very large eccentricities (15–20°) are also typically not only fixated by an eye movement but by a combined head and eye movement (Sanders, 1970; Stark et al., 1975, but see also Oommen and Stahl, 2005).

When: Saccade Reaction Time

When the brain initiates a saccade can also be regarded from two different perspectives. In extended, complex behaviors, several saccades and fixations are made in sequence. The temporal aspects of saccade generation have been examined from the perspective of the fixation by measuring fixation durations (reviewed by Liversedge and Findlay, 2000; Rayner, 2009). Of interest to this thesis is a second perspective, which considers the processing that can be ascribed to saccade preparation rather than processing during the fixation period. To investigate saccades from this perspective, studies have used target onsets to elicit saccades and measure the saccade reaction time (SRT). This is the time from the saccade target's onset to the initiation of the saccade (Fig. 3 B). Measurements of SRTs can be used to probe which information is used for saccadic control, for example, to examine how the location or visual properties of the stimulus (e.g., color or luminance) affect saccade initiation (e.g., Kalesnykas and Hallett, 1994; Pelisson and Prablanc, 1988; Reuter-Lorenz et al., 1991).

Repetitions of many SRT measurements can be used to obtain a distribution of SRTs (Fig. 4 B). The mode of this distribution is located at approximately 200 ms. Bimodal distributions are also observed in the gap task (Saslow, 1967). In this task, the fixation stimulus disappears shortly before the onset of the saccade target. In this case, a second mode typically occurs at around 150-180 ms. Saccades in this SRT range are attributed to a separate class of saccades, which are referred to as *express saccades* (Fischer et al., 1993; but see Wenban-Smith and Findlay, 1991; Cavegn and Biscaldi, 1996).

How: Saccade Kinematics

During a saccade, the eye achieves velocities of up to $500\,°/s$ (Fuchs, 1967; Leigh and Zee, 2006). Because of this, vision is severely affected during the movement. Some writers argue that vision is also selectively suppressed during a saccade, a phenomenon that is known as *saccadic suppression* (Dodge, 1900; Zuber and Stark, 1966; reviewed by Ross et al., 2001).

The trajectories of saccades are very stereotypical. With few exceptions (e.g., McPeek et al., 2003), the saccade moves the eye along the shortest path in a straight line to the new location. The velocity and duration primarily depend on the distance to the target and the required saccade amplitude. The relationship between saccade duration and amplitude, and peak velocity and amplitude is referred to as the *main sequence* (Bahill et al., 1975; Baloh et al., 1975; Lebedev et al., 1996). Fig. 4 A shows the typical distribution of measured velocities as a function of the saccade amplitude. Similar to SRT measurements, the main sequence, and in particular deviations from it, are used to make

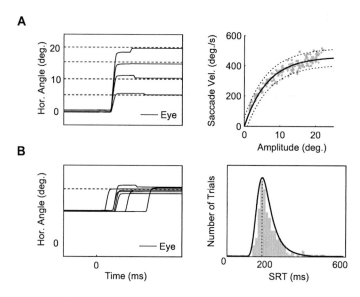

Figure 4: **A** Left: Horizontal saccades to targets at varying eccentricity (broken lines) aligned at saccade onset. Right: Scatterplot of saccade peak velocities showing the exponential relationship between saccade amplitude and peak velocity. **B** Left: Horizontal saccades to a target at a single eccentricity aligned at target onset. Right: Histogram showing the typical unimodal distribution of saccade reaction times (SRTs) that can be observed in saccades following the onset of a peripheral target (adapted from Bieg et al., 2012).

statements about the factors that influence saccadic control (Schmidt et al., 1979; Jürgens et al., 1981; Guan et al., 2005).

Neurophysiology

How does the brain generate saccades? What we know about saccade generation stems from studies that investigated saccade characteristics (see previous chapter) and from neurophysiological studies. This section highlights some aspects of saccade generation that are of particular importance to this thesis. For more comprehensive reviews, the reader is referred to Carpenter (1988); Moschovakis et al. (1996); Schall and Thompson (1999); Sparks (2002); Scudder et al. (2002); Glimcher (2003); Girard and Berthoz (2005); Krauzlis (2005); Leigh and Zee (2006); Schall and Boucher (2007); Gold and Shadlen (2007); Bisley and Goldberg (2010); Gandhi and Katnani (2011); Schall et al. (2011); Paré and Dorris (2011).

The motor signals that control saccadic eye movements are generated by nuclei in the *brainstem* (Fig. 5). The eye's six extraocular muscles are innervated by three cranial nerves (III, IV, VI). The control signal of these nerves specifies both the desired velocity during the saccade and final position of the eye after the saccade. The saccade is accelerated by a pulse of rapid firing by premotor burst neurons (Robinson, 1964; Fuchs and Luschei, 1970). The intensity of the pulse controls the saccade's velocity. Its duration determines the saccade's duration and, hence, its amplitude. A tonic step signal that follows the pulse is necessary to keep the eye in its eccentric position (Cannon and Robinson, 1987). During fixation, premotor burst neurons are continuously inhibited by omnipause neurons. Together with the activation of burst neurons, inhibition of omnipause neurons is a precondition to saccade initiation (van Gisbergen et al., 1981).

The mechanisms in the brainstem are relatively well understood. However, several other areas in the brain are also involved in saccadic control. Three major areas are the superior colliculus, frontal eye fields, and lateral intraparietal area (Fig. 5).

The *superior colliculus* (SC), a nucleus at the roof of the brainstem, is most closely related to the activity in the oculomotor nuclei and relays most of the information from cortical areas. The SC's activity

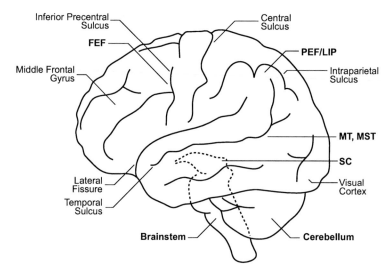

Figure 5: Brain areas that are involved in saccade generation and eye movements in general (in bold): Brainstem, Cerebellum, Superior Colliculus (SC), Frontal Eye Fields (FEF), Parietal Eye Fields (PEF), human homologue of the Lateral Intraparietal Area (LIP), Middle Temporal Area (MT), Middle Superior Temporal Area (MST). Anatomic information from Duvernoy (2007) and others (see main text).

likely results in the characteristic saccade main sequence (see previous chapter and Goossens and Van Opstal, 2006; Harris and Wolpert, 1998; but also Quaia et al., 1999). As all of the areas that are part of the oculomotor network, the SC's activity is not only related to eye movements. Specifically, the SC is thought to implement a more general orienting system that controls body, head, and eye movements (Freedman et al., 1996). Furthermore, the SC also exhibits activity that is related only to vision (Cynader and Berman, 1972) and hand movements (Werner, 1993).

An area that is closely related to the oculomotor activity of the SC are the *frontal eye fields* (FEF). They are connected to other cortical areas and project directly to the brainstem and SC (Segraves and Goldberg, 1987). This area is thought to be a main relay station for cortically driven, voluntary eye movements (Schiller et al., 1980; Hanes and Wurtz, 2001). The FEF are not only implicated in oculomotor control but are also engaged in covert attention (Moore and Fallah, 2001; Moore and Armstrong, 2003).

The third major area is the *lateral intraparietal area* (LIP). Area LIP receives inputs from a several cortical sites, most prominently, extrastriate areas, and projects to FEF and SC. Early results on parietal activity have lead researchers to interpret LIP as a parietal eye field (PEF), in analogy to the activity in FEF (Andersen et al., 1992). However later work has confirmed that LIP, even more so than FEF, is likely an area that is primarily concerned with visual attention rather than eye movements per se (reviewed by Paré and Dorris, 2011).

Other notable areas that are related to saccadic behavior are the supplementary eye fields and prefrontal cortex (Munoz and Everling, 2004), thalamus (Sommer and Wurtz, 2008), basal ganglia (Hikosaka et al., 2000), and cerebellum (Scudder et al., 2002). For further details on these areas the reader is referred to the provided references.

Generation Principles

In comparison to the oculomotor nuclei in the brainstem, the function and coordination of these extended brain areas is less clear (Girard and Berthoz, 2005). Converging evidence suggests that the main role of these areas, FEF and SC in particular, is to (1) maintain a representation of potential saccade goals and (2) select one of these targets, to eventually trigger a saccade by supplying the brainstem generator with its crucial inputs: the desired location of the saccade (where) and the trigger impulse (when).

More generally, the brain regions that are part of the oculomotor network could be considered to implement an orienting system that is based on a salience or *priority map* (Itti and Koch, 2001; Fecteau and Munoz, 2006; Bisley et al., 2011). This map is used to keep track of important locations in the environment and serves as the basis for the selection process. Locations of the map that are selected by this process receive preferential processing, either by a covert shift of attention or an eye movement.

The neural organization that can be observed in all three areas (SC, FEF, LIP) support this idea. Neurons in intermediate layers of SC are *topographically* organized. The direction and amplitude of saccades is systematically linked to this organization (Robinson, 1972). Topographic representations can also be found in FEF (Bruce et al., 1985) and LIP (Sereno et al., 2001; Steenrod et al., 2013).

Neurons or pools of neurons on these maps implement a selection process (Dorris et al., 1997; Hanes and Schall, 1996; Shadlen and Newsome, 2001). For example, neurons in FEF increase their activity shortly before a saccade (Hanes et al., 1995). The FEF neuron firing rate before the saccade onset is relatively similar for each saccade (but see also Jantz et al., 2013). However, the rate of rise in the activity is not and correlates well with SRTs (Hanes and Schall, 1996; Hanes et al., 1998). This suggests that the neuronal activity is directly linked to saccade initiation and has been modeled by assuming a *competition* between neuron populations that represent different eye movement goals (Trappenberg et al., 2001; McPeek et al., 2003). The neuron or neuron population that wins this race then determines the endpoint of the next saccade in a winner-take-all fashion (Itti and Koch, 2001).

Race models have been proposed before to explain the variability of reaction times in general (Logan et al., 1984) and SRTs in particular (Carpenter, 1988; Carpenter and Williams, 1995). These models assume that a response is elicited after a decision signal rises over time to a threshold, for instance, a certain neuron firing rate. The decision signal is often related to the amount of information that can be accumulated concerning the relevant decision, for example, the sensory evidence about the presence or absence of a target. The nature of this accumulation process is stochastic because of the perceptual limitations of the sensory apparatus that supplies this information (eye, visual cortex, etc.). Support for this idea is provided by the neurophysiological recordings mentioned above and behavioral findings that explain changes in the distribution of SRTs based on such a model. For example, increasing the amount of information that is available to observers about the correct saccade target leads to an increase in the rate of rise (Reddi et al., 2003), prior information about the correct location leads to a bias which shows as a lowering of the saccadic threshold (Carpenter and Williams, 1995).

To summarize, behavioral and neurophysiological findings support the idea that a salience or priority map underlies the control of saccadic eye movements. This map integrates information about the targets in the environment and its activity determines the characteristics of saccades, for example, their landing position or onset time.

1.5 SMOOTH PURSUIT

Smooth pursuit eye movements constitute another class of eye movements. In comparison to saccades, smooth pursuit eye movements are much slower (typically below $100°/s$). The function of these slow movements complements that of saccades. While saccades reorient the foveal region to a new location in the visual field, smooth pursuit eye movements keep the fixated location within the fovea to assure optimal vision also when the object moves through the visual field. The following sections present fundamental characteristics of smooth pursuit. For a more detailed description the reader is referred to Lisberger et al. (1987); Ilg (1997); Krauzlis (2005); Orban de Xivry and Lefèvre (2007); Barnes (2008, 2011).

Where: Target and Motion Direction

Where smooth pursuit moves the eye is closely linked to which target is the object of pursuit. Smooth pursuit eye movements are usually considered to be voluntary, despite the fact that they cannot be elicited for extended time periods without a smoothly moving visual target (with exceptions, e.g., Gauthier and Hofferer, 1976; Berryhill et al., 2006). However, observers can exert voluntary control over which target is to be pursued — a selection process that is quite similar to the one in saccades (Gardner and Lisberger, 2001; Case and Ferrera, 2007).

The precise direction of smooth pursuit is determined by the movement direction of the selected target. Typically, the eye is moved in the direction of the target's motion, even if this momentarily moves the target farther away from the fovea (Lisberger and Westbrook, 1985, but see also Blohm et al., 2005)

When: Pursuit Latency

The onset of smooth pursuit can be studied in the laboratory using specialized experimental tasks. A widely used procedure is the presentation of a *ramp stimulus*. Here, a static object suddenly begins to move with constant velocity in one direction. The resulting behavior shows that smooth pursuit begins after a latency period of ca. 100 ms (Wyatt and Pola, 1987).

Because of this reaction time, a typical sequence of smooth pursuit initiation is as follows (Fig. 6, see also Lisberger, 1998): After the initial latency the eye is briefly accelerated in the pursuit target's motion direction. At this point, the velocity of the eye typically does not match that of the target. Due to the latency, the retinal error to the target accumulated by this time. Therefore, an initial catch-up saccade is usually triggered after ca. 200 ms that foveates the retinal image of the target. After this, pursuit velocity increases to match that of the target and the eye tracks the target smoothly (steady-state).

The *Rashbass* paradigm (Rashbass, 1961) is an experimental paradigm that is used to study smooth pursuit without the initial saccade, whose occurrence may be detrimental to an exact analysis of the motion properties in the first 100-200 ms. In this paradigm, the target undergoes a step-ramp such that it moves across its original location within the typical latency of the initial saccade (i.e., 200 ms). From the perspective of the observer, the image of this target moves *foveopetally*, that is, from a peripheral location toward the fovea. Under these conditions, smooth pursuit is frequently initiated without the initial saccade or the saccade is canceled and re-planned, prolonging its onset until after the target crosses its original location and thus the observer's fovea (Gellman and Carl, 1991).

How: Catch-Up Saccades

Smooth pursuit gain is the ratio of the target velocity and eye velocity and is typically used to describe how well the eye matches the target motion. Pursuit of single targets that move in a straight line with constant velocity can typically be achieved with a gain close to one. For target speeds higher than $100°/s$, gain deteriorates (Meyer et al., 1985). As a consequence, the eye falls behind. The resulting retinal error is then compensated for by small catch-up saccades.

Neurophysiology

Smooth pursuit eye movements invoke a similarly complex network of brain areas as saccades. In fact, evidence now exists that both types of eye movements partially rely on the same network of subsystems (reviewed by Krauzlis, 2005; Orban de Xivry and Lefèvre, 2007). This section briefly highlights two additional areas that are of particular importance to smooth pursuit, namely the medial temporal area and the medial superior temporal area. More extensive reviews of these and other brain areas that control smooth pursuit are presented by Lisberger et al. (1987); Ilg (1997); Krauzlis (2005); Thier and Ilg (2005); Büttner and Kremmyda (2007); Orban de Xivry and Lefèvre (2007); Ilg (2008); Barnes (2008, 2011).

The importance of the *medial temporal area* (MT) (Zeki et al., 1991; Tootell et al., 1995; Tanabe et al., 2002) in smooth pursuit generation has been demonstrated by lesioning studies that spared saccadic but impaired smooth pursuit behavior (Newsome et al., 1985). MT is located close to the primary visual cortex and receives projections from motion-sensitive cells that are located there (Fig. 5). MT is considered to not only subserve smooth pursuit but motion processing in general (reviewed by Born and Bradley, 2005; Ilg, 2008).

The *medial superior temporal area* (MST) (Dukelow et al., 2001; Tanabe et al., 2002) receives inputs from area MT and is primarily known for

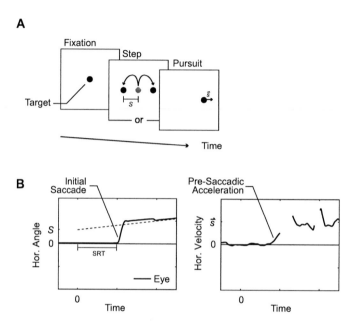

Figure 6: **A** Laboratory smooth pursuit task. The observer fixates on a central fixation object. After a random foreperiod, the target jumps to eccentricity S where it begins to move peripherally with constant velocity $\vec{s}$. **B** Typical response to such a step-ramp motion. An acceleration of the eye in the direction of the pursuit target's motion can be observed shortly before the initial saccade moves the eyes to the target. The pursuit velocity then increases to match that of the target's. Saccadic periods are omitted in the velocity plot.

its role in the optic flow computations that are necessary for self-motion (reviewed by Andersen et al., 1997). However, it also supplies a specialized motion signal for smooth pursuit. For example, pursuit-related neurons in this area are not only sensitive to retinal but also illusory motion (Ilg and Thier, 2003). Furthermore, its computations of optic flow may also play a role in dealing with problems of self-induced image motion (reafference) in smooth pursuit (Lindner and Ilg, 2006).

Information from these two cortical areas is relayed by nuclei in the pons to the cerebellum. From here, the smooth pursuit signal projects via the vestibular nucleus to the extraocular motor neurons.

Generation Principles

Smooth pursuit faces a similar problem as saccades, namely the selection of a target. Selection of the smooth pursuit target seems to rely on the same principle as in saccades, namely *competition* between represented motion directions. This is reflected in the early, open-loop phase of pursuit. For example, if multiple targets are presented, this early acceleration often reflects a weighted average of the presented target motions (Ferrera and Lisberger, 1995). However, after the initial saccade, the velocity typically matches exactly one of the targets, similar to the winner-take-all schema that is employed in the selection of saccade targets. Interestingly, post-saccadic pursuit velocities still reflect the average of two targets in those cases in which the saccade endpoint also reflects the average of both target locations (Gardner and Lisberger, 2001, see also Spering et al., 2006). Smooth pursuit is also affected by the offset of the fixation stimulus in the gap paradigm. Similar to saccade RTs, the onset of pursuit occurs earlier when the fixation stimulus disappears before the onset of the target stimulus (Krauzlis and Miles, 1996). This evidence highlights the similarities between both types of eye movements and suggests that the selection and initiation process may rely on the same mechanisms (Kornylo et al., 2003; Krauzlis, 2003).

Unlike in saccades, visual information is available during the movement for on-line corrections of the smooth pursuit velocity. In contrast to saccadic control, this suggests a simple closed-loop architecture for smooth pursuit generation. However, due to visual processing delays, an accurate response can only be generated by a system that develops an internal *prediction* of the target's motion (Bahill and McDonald, 1983; Robinson et al., 1986).

Further evidence for an internal model of the target's motion is provided by studies that showed continuation of smooth pursuit despite temporary blanking of the pursuit target (Becker and Fuchs, 1985; Bennett et al., 2007) or tracking of illusory targets (Ilg and Thier, 2003). The level of refinement of this internal model likely depends on the duration that observers are exposed to the pursuit target's motion prior to pursuit and on previous experience (Deno et al., 1995; Bennett et al., 2007). For example, it can be shown that observers are able to begin smooth pursuit in anticipation of the target's onset based on their experience from previous trials (Kao and Morrow, 1994; Freyberg and Ilg, 2008).

It should be noted that prediction of a target's motion not only plays a role in pursuit but also in saccade planning. For example, it has been shown that saccades to moving stimuli typically take into account the accumulation of retinal error during the saccade preparation period such that the saccade lands directly on the target (Keller and Johnsen, 1990; Gellman and Carl, 1991; Eggert et al., 2005b; Etchells and Benton, 2010).

In summary, smooth pursuit and saccadic eye movements show similarities in their organization principles, presumably reflecting the fact that they typically co-occur to complement each other (Orban de Xivry and Lefèvre, 2007).

1.6 THESIS OVERVIEW AND DISCUSSION

It is still a mystery how the brain organizes its eye movement repertoire so efficiently and elegantly that we hardly notice the visual limitations of our eyes (Section 1.1). The secret of the brain's organization talent lies in its ability to coordinate eye movements according to task demands (Land and Tatler, 2009). How is this coordination achieved, for example, in common behaviors such as reaching and grasping (Land and Hayhoe, 2001)? Moreover, the brain can choose between different classes of eye movements, most notably, between a saccadic or smooth pursuit eye movements (Sections 1.4 and 1.5). What are the principles that govern this choice? This thesis summarizes a series of experiments in human observers that help us to address these questions.

The presented studies can be topically organized in two parts: Part one, which comprises Chapter 2, is related to work on the coordination of saccadic eye movements and hand movements. Part two, which includes Chapters 3 and 4, concerns the coordination of saccades and smooth pursuit eye movements. The following sections summarize and discuss each part and chapter.

Part 1: Coordination of Eye and Hand Movements

Eye movements are tightly coordinated with hand movements. This can be observed in everyday situations: For example, the eyes typically lead actions to locate objects and guide the hand (Land and Hayhoe, 2001). The same observation can be made in the laboratory (Bekkering et al., 1994) among other phenomena: Pointing movements are less accurate if the target is not fixated (Prablanc et al., 1979) and gaze cannot easily be removed from target fixation once a pointing movement is in progress (Neggers and Bekkering, 2000).

Several studies also found that the characteristics of saccadic eye movements appear to be influenced by concomitant hand movements. For example, saccades during pointing movements have shorter SRTs (Lünenburger et al., 2000; Lünenburger and Hoffmann, 2003), exhibit higher peak velocities, and lie on a different main sequence (Epelboim et al., 1997; Snyder et al., 2002). Do these phenomena reflect mechanisms that coordinate eye and hand movements (Lünenburger et al., 2000)?

In *Chapter 2* of this thesis we argue against this view. A problem in studies of eye-hand coordination arises if the usefulness of the visual information that is available through the saccadic eye movement is not balanced between the tested conditions. For example, in a visually guided pointing task, a saccade to the target provides foveal vision, which is used to adjust the finger movement. On the contrary, visual information in the control condition, which requires observers to simply look at the target as quickly as possible, does not support an accompanying task (Epelboim et al., 1997). We argue that the characteristics of saccades supporting hand movements may not be influenced by the coordinated motor action, but by the visual requirements of the task that they contribute to.

This hypothesis is supported by our experimental results. Saccade RTs and velocities were measured in a step-movement task in which observers fixated a target as quickly as possible. These measurements were compared with those in a discrimination task. In this task, the saccade to the target was necessary to discriminate an object. The comparison shows a similar reduction in SRTs and increase in velocities as those observed in the studies on eye-hand coordination (e.g., Epelboim et al., 1997). This suggests that concurrent hand actions may not necessarily influence saccades directly but indirectly. Chapter 1.4 outlines how saccade-related areas in the brain maintain a map of potential saccade locations. The need to fixate the target in order to extract task relevant information may have imbued the target's representation in such a map with a higher activation value. This elevation may in turn explain the measured differences in saccade characteristics.

Evidence for this view also comes from studies that explore the relationship between reward value and saccadic behavior. For example, monkeys performed saccades to target locations that promised higher juice rewards with shorter latency and higher velocities (Kawagoe et al., 1998). Human SRTs are influenced by both the expected monetary reward value and probability (Milstein and Dorris, 2007) and show increased sensitivity to oculomotor capture by stimuli that were previously associated with monetary reward (Theeuwes and Belopolsky, 2012). The underlying neurophysiological mechanisms for these phenomena may be found in the modulatory connections of the basal ganglia to saccade-related areas such as the superior colliculus (reviewed by Hikosaka et al., 2000).

The links to ideas about the representation of saccade goals by the brain are speculative. As Jeffrey Schall rightfully pointed out, behaviors that are superficially similar may stem from entirely different neuronal processes (Schall, 2004). A definitive answer to the representations of saccades during coordinated hand actions can therefore only be provided by studies that also monitor the activity in the brain during these actions.

Part two of this thesis focuses on the coordination of saccades and smooth pursuit eye movements. In many situations outside the laboratory, the fixated stimulus is not static but moves through the visual field. In these situations, foveation of the target is achieved by a combination of saccadic and smooth pursuit eye movements. To understand how the brain controls saccades, it is therefore also important to understand how they are coordinated with smooth pursuit eye movements.

The processes that trigger saccades and smooth pursuit are closely related (Orban de Xivry and Lefèvre, 2007). Yet, the nature of this relationship is still unclear. Some experimental manipulations affect the onset timing of both eye movements, for example, offsets of the fixation stimulus in the gap paradigm (see Chapters 1.4 and 1.5). Other studies show mutual interactions. For instance, the latencies of saccades that terminate pursuit and move the eyes to a new target are *asymmetric*: SRTs are shorter if the target appears congruently, that is, in the direction of pursuit rather than in the opposite direction (e.g., Tanaka et al., 1998; Khan et al., 2010). A plausible explanation for this phenomenon suggests that saccade initiation in the case of congruent saccades benefits from a broad bias of visual attention, which may have been caused by the pursuit movement (van Donkelaar and Drew, 2002).

This bias in attention may be reflected in a second phenomenon, namely the propensity to perform anticipatory saccades in the direction of pursuit. This tendency depends on the nature of the smooth pursuit task. For example, pursuit is smoother and fewer anticipatory saccades can be observed when the pursuit task is coupled with a visual task (Van Gelder et al., 1990). This phenomenon has been linked to an interaction between the observer's attention and the automatic anticipation processes that are a hallmark of pursuit (Chapter 1.5). In other words, anticipation might be exaggerated because of the observer's conscious effort to pursue the target as precisely as possible (Van Gelder et al., 1990, 1995a). Asymmetries in SRTs may be the result of this exaggeration and may therefore constitute a phenomenon that would not be present during more natural (i.e., functional) pursuit conditions.

Chapter 3 examines this assumption. Our experimental results show that asymmetries in SRTs can also be observed when pursuit subserves a perceptual function. In the experimental test condition, participants' pursuit eye movements were recorded while they followed the pursuit target by steering an on-screen cursor. In the control condition, participants only followed the pursuit target by eye. The results show improved smooth pursuit eye movements with fewer anticipatory saccades that interrupt pursuit in the test condition. Yet, saccades to a peripheral target were still initiated earlier in this condition, when they moved the eye in the direction of pursuit. This provides evidence against the hypothesis that SRT asymmetries are a by-product of laboratory pursuit task conditions and emphasizes the robustness of this phenomenon.

Asymmetries in SRTs do not only occur in saccades *from* pursuit. The results presented in *Chapter 3* show that SRTs of saccades *to* pursuit also depend on the relative motion direction of the pursuit target. In both conditions, SRTs in the direction of pursuit are shorter. We argue that the explanations that were put forth to account for the asymmetries in saccades from pursuit (e.g., Khan et al., 2010; van Donkelaar, 1999) cannot simply be applied to saccades to pursuit. Instead, asymmetric SRTs in saccades to a moving target may be related to the mechanisms that select the appropriate oculomotor response (saccadic or smooth). In foveopetal movements, that is, movements of the pursuit target toward the currently fixated location, and under certain conditions, smooth pursuit can commence without an initial saccade (see Chapter 1.5 and Rashbass, 1961). In these trials, some authors also report longer SRTs when an initial saccade does occur (Gellman and Carl, 1991; Moschner et al., 1999). However, the exact conditions that lead to these delays are unclear.

Chapter 4 presents results that enable us to characterize these conditions in greater detail. Here, the SRTs of saccades to objects that either move toward (foveopetal) or away from the observer's fovea (foveofugal) are compared in different speed and step-amplitude conditions. The results demonstrate that SRTs of saccades at the onset of pursuit depend on the motion direction of the pursuit target: SRTs in foveopetal trials are longer than those in foveofugal trials. Moreover, the results also provide evidence that the SRTs in foveopetal trials depend on a combination of target eccentricity and speed. The same dependency on the target's eccentricity and speed also determines whether a saccade generally occurs at the onset of foveopetal trials or whether smooth pursuit is initiated directly (see Chapter 1.5 and Rashbass, 1961; de Brouwer et al., 2002b). These results point to a similarity between the processes that determine the type of oculomotor response and the onset time of a saccade. The following section expands the discussion in *Chapter 4*, to explain the neurophysiological mechanisms that are assumed to give rise to both phenomena.

Neurophysiology of Saccade-Pursuit Interaction

A physiologically plausible mechanism that governs the decision between a saccadic and smooth response is suggested by Grossberg et al. (2012). This circuitry suppresses one of the response options, namely the saccade, once the target's neuronal representation enters a foveal zone. Their model assumes a cortico-tectal pathway that engages two mechanisms in the brainstem that prevent saccade initiation (Fig. 7): First, fixation neurons in the superior colliculus (SC) activate omnipause neurons (OPNs) in the brainstem's reticular formation. OPNs inhibit the burst neurons that provide the saccadic motor command (Scudder et al., 2002). Second, SC fixation neurons locally inhibit SC movement neurons and thereby directly affect the source of the saccadic drive signal that is sent to the burst neurons (Munoz and Istvan, 1998; Paré and Hanes, 2003). These two mechanisms are engaged by signals from foveal locations of the motion-sensitive medial temporal (MT) area, which are thought to influence the corresponding fixation neurons in the SC (Fries, 1984; Collins et al., 2005; Grossberg et al., 2012).

In *Chapter 4*, SRTs to targets that move foveopetally are shown to be prolonged. This increase in saccadic latency can be regarded as a by-product of the cortico-tectal suppression mechanism outlined by Grossberg et al. (2012). The activity of SC movement neurons before saccade onset is related to SRTs (Dorris et al., 1997; Dorris and Munoz, 1998). Inhibitory or excitatory input to these neurons not only determines whether a saccade will be generally initiated but also regulates when it occurs. Local inhibitory connections from fixation neurons that are activated through the cortico-tectal pathway

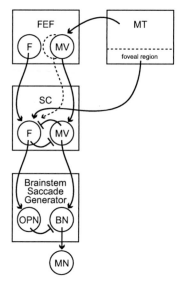

Figure 7: Potential pathways that determine suppression of saccades at the onset of smooth pursuit. See text for details. Frontal Eye Fields (FEF), Superior Colliculus (SC), Fixation Neuron (F), Movement Neuron (MV), Omnipause Neuron (OPN), Burst Neuron (BN), Motor Neuron (MN). Broken circle in FEF indicates a potential receptive field response shift by motion information (Cassanello et al., 2008).

could result in prolonged SRTs in instances, wherein the inhibition of movement neurons is not strong enough to prevent the saccade entirely but sufficient to retard the buildup of activity in these cells. SRT asymmetries could be the result of differences in the inhibition of movement neurons in the SC. During foveopetal movements, a target's neuronal representation moves toward the foveal zone in area MT, which increases the inhibition of movement neurons. The target's representation moves away from the foveal zone during foveofugal movements, releasing the inhibition of movement neurons. In addition, the asymmetry in SRTs may be caused by asymmetries in the motion-sensitive area itself. For example, the results discussed in *Chapter 4* also show a stronger pre-saccadic pursuit response during foveopetal movements. This observation can be attributed to asymmetries in the output of motion sensitive areas such as MT (Lisberger and Westbrook, 1985; Tychsen and Lisberger, 1986 and discussion in *Chapter 4*).

The direct connection between MT and SC as proposed by Grossberg et al. (2012) is not the only neuronal route by which motion information could influence saccade onsets. Earlier investigators have argued that the decision between a saccadic and smooth response relies on an extrapolation of the target's position based on its velocity (see Chapter 1.5 and Gellman and Carl, 1991; de Brouwer et al., 2002b). The neuronal connections outlined thus far do not implement a mechanism that extrapolates the target's position. Yet, saccades to moving stimuli are accurate and take into account movements during the saccadic latency period (Keller and Johnsen, 1990; de Brouwer et al., 2002a; Guan et al., 2005). This illustrates that extrapolation exists in the oculomotor system, which adjusts the amplitude of a saccade. A mechanism that implements these computations was recently suggested by Cassanello et al. (2008). Motion information from area MT is not only relayed to the SC but also influences a second cortical area that is related to eye movements, namely the frontal eye fields (FEF, Schall et al., 1995). According to Cassanello et al. (2008), motion information that is sent to FEF scales the response of FEF neurons according to the retinal velocity of the target. This effectively shifts the response characteristics of FEF neuron populations in the direction of motion. For example, responses shift toward the foveal area in the FEF topography in the case of foveopetal target movements. This shift in FEF responses can be regarded as an example for extrapo-

This thesis comprises a collection of manuscripts that are either published or prepared for publication. Details about these manuscripts are presented in the following.

The ideas for the studies, their experimental design and software were developed by the candidate. The data collection and analysis was performed by the candidate. The co-authors supervised the work of the candidate and assisted in the revision of the manuscripts.

- Bieg, H.-J., Bresciani, J.-P., Bülthoff, H. H., & Chuang, L. L. (2012). Looking for Discriminating is Different from Looking for Looking's Sake. *PLoS ONE*, 7(9), e45445.

- Bieg, H.-J., Bresciani, J.-P., Bülthoff, H. H., & Chuang, L. L. (2013). Saccade reaction time asymmetries during task-switching in pursuit tracking. *Experimental Brain Research*, 230(3), pp. 271-281.

- Bieg H.-J., Chuang, L. L., Bülthoff, H. H., & Bresciani, J.-P. (2013). Asymmetries in saccade reaction times to pursuit (submitted).

Parts of this work were also presented at the following conferences:

- Bieg H.-J., Chuang L. L. and Bülthoff H. H. (2010): Does adding a visual task component affect fixation accuracy?, 33rd European Conference on Visual Perception, Lausanne, Switzerland, *Perception*, 39(ECVP Abstract Supplement) 35.

- Bieg H.-J., Bresciani J.-P. , Bülthoff H. H. & Chuang L. L. (2012): Asymmetries in saccadic latencies during interrupted ocular pursuit, 35th European Conference on Visual Perception, Alghero, Italy, *Perception*, 41(ECVP Abstract Supplement) 137.

- Bieg, H.-J., Bülthoff, H., & Chuang, L. (2013). Attentional Biases during Steering Behavior. (V. G. Duffy, Ed.) Digital Human Modeling and Applications in Health, Safety, Ergonomics, and Risk Management. Healthcare and Safety of the Environment and Transport (pp. 21-27). Berlin: Springer. Lecture Notes in Computer Science Volume 8025.

- Bieg H.-J., Chuang L. L., Bresciani J.-P., Bülthoff H. H. (2013) Asymmetric saccade initiation at smooth pursuit onset. *23rd Oculomotor Meeting*, Linz, Austria.

2

COORDINATION OF EYE AND HAND MOVEMENTS

This chapter has been reproduced from an article published in PLoS ONE: Bieg, H.-J., Bresciani, J.-P., Bülthoff, H. H., & Chuang, L. L. (2012). Looking for Discriminating is Different from Looking for Looking's Sake. *PLoS ONE*, 7(9), e45445.

2.1 ABSTRACT

Recent studies provide evidence for task-specific influences on saccadic eye movements. For instance, saccades exhibit higher peak velocity when the task requires coordinating eye and hand movements. The current study shows that the need to process task-relevant visual information at the saccade endpoint can be, in itself, sufficient to cause such effects. In this study, participants performed a visual discrimination task which required a saccade for successful completion. We compared the characteristics of these task-related saccades to those of classical target-elicited saccades, which required participants to fixate a visual target without performing a discrimination task. The results show that task-related saccades are faster and initiated earlier than target-elicited saccades. Differences between both saccade types are also noted in their saccade reaction time distributions and their main sequences, i.e., the relationship between saccade velocity, duration, and amplitude.

2.2 INTRODUCTION

Saccades are rapid eye movements which are performed 3–4 times a second to fixate on a different spot in the environment (Findlay and Gilchrist, 2003). The characteristics of saccades, notably *target-elicited saccades*, which follow the onset of a visual stimulus, have been thoroughly investigated. Past research has explored how visual properties of the saccade target, for instance, its luminance, color, or spatial arrangement, influence saccade planning and execution. For example, brighter stimuli lead to quicker initiation of saccades (Reuter-Lorenz et al., 1991). In experiments such as this, the saccade is elicited by the appearing target but the task does not inherently require the participant to fixate. This contrasts with the situation outside the laboratory. Here, saccades redirect the fovea, the region with highest visual acuity on the retina, to perform specific visual tasks (Land, 2009). The purpose of this paper is to compare the characteristics of classical *target-elicited saccades*, which do not require fixation per se, to *task-related saccades*, which require fixation due to task demands. Considering this distinction is important to avoid potential confounds in experimental tests of the oculomtor system's variability.

The functional variability of saccade properties has been the topic of previous work, in particular work related to visually guided motor actions. Visual information is critical for accurate grasping and pointing (Abrams et al., 1990; Land et al., 1999; Neggers and Bekkering, 2000; Binsted et al., 2001). The need to coordinate eye and hand movements could therefore be one factor that influences saccade characteristics. To test this hypothesis, Epelboim et al. (1997) measured differences in saccade velocities across two conditions. One condition required participants to fixate a sequence of targets and the other to tap on them with a finger. Tapping resulted in faster saccades and a change in the relationship between saccade velocity, duration, and amplitude. This relationship, which is referred to as the saccadic *main sequence*, was thought to be the stereotypical result of brainstem saccade generator mechanics (Bahill et al., 1975). The work by Epelboim and colleagues demonstrates that changes in the main sequence occur when participants are engaged in an oculomanual task such as pointing. In a similar study with monkeys, Snyder et al. (2002) found higher peak velocities and shorter durations for saccades that accompany arm movements. Like Epelboim et al., Snyder and colleagues also report main sequence differences. Apart from changes in saccade velocity, other studies reported differences in saccadic reaction time (RT), the time required to initiate a saccade following stimulus onset. For example, Lünenburger and colleagues (Lünenburger et al., 2000; Lünenburger and Hoffmann, 2003) found that saccades that support rapid pointing movements are initiated earlier than saccades that are made without such a movement to the target. In explaining their findings, Lünenburger et al. (2000) suggested that saccade reaction times are adjusted to synchronize eye fixation so that foveal vision is provided during the final phase of the pointing movement.

This body of research suggests a functional role of saccade property adjustments due to the need to coordinate vision and hand movements. But are such adjustments only specific to oculomanual coordination? A study by Montagnini and Chelazzi (2005) casts doubt on this assumption. In their study participants were not engaged in an oculomanual task but were required to rapidly identify an alphabetic letter at the saccade endpoint. Their results show similar changes in saccade properties, namely a decrease in saccade reaction time and an increase in velocity, when participants performed the rapid identification task in comparison to a condition where they only looked at the targets in succession. Related to this is the finding that saccades can be altered by verbally instructing participants to either emphasize speed or accuracy (Reddi and Carpenter, 2000). A comparison of differences in saccade RT distributions that were observed in this study with those observed in the study by Montagnini and Chelazzi suggests that the underlying process that leads to the reduction of RTs when performing an identification task could be different from the process that leads to the RT reduction when participants receive verbal instruction to emphasize speed over accuracy (Montagnini and Chelazzi, 2005). Instead of assuming a general effect of time pressure as it might be induced by verbal instructions, Montagnini and Chelazzi (2005) therefore explained their findings on the grounds of *perceptual urgency*, i.e., as a natural response of the oculomotor system to stimuli that are only available very briefly.

The need to rapidly process visual information at the saccadic endpoint may cause changes in saccade properties (Montagnini and Chelazzi, 2005). This could also be an explanation for the results that were observed in the previously cited studies on oculomanual coordination. For example, differences in saccade characteristics in the studies by Epelboim et al. (1997) and Snyder et al. (2002) could stem from the need to perform two concurrent motor acts (eye and hand movements) or from the fact that movements of the eyes served a perceptual purpose in one but not in the other condition.

Two different studies provide additional support for the idea that saccades might show different characteristics if they are followed by a perceptual task. In experiments similar to that of Montagnini and Chelazzi (2005) by Trottier and Pratt (2005) and Guyader et al. (2010), lower saccade RTs were measured for saccades that supported a visual discrimination task. It is important to note that, unlike in the study by Montagnini and Chelazzi (2005), time pressure was not explicitly induced during these experiments.

The picture that emerges from this body of research suggests a general difference between classical target-elicited saccades and task-related saccades; a difference which might have confounded previous studies on motor coordination (Epelboim et al., 1997; Snyder et al., 2002; Lünenburger et al., 2000; Lünenburger and Hoffmann, 2003) or time pressure (Montagnini and Chelazzi, 2005). To test this, the current study compared classical target-elicited saccades to saccades that supported a visual discrimination task. Experiment 1 compared saccade RTs, peak velocity, duration, and gain in both types of saccades. Specifically, differences in the distribution of saccade RTs were measured. This made it possible to examine theoretical influences on saccade generation and enabled a comparison with the study by Montagnini and Chelazzi (2005). In experiment 2, saccade velocities and duration were measured across a range of amplitudes for both saccade types. This data was used to establish the velocity and duration main sequence. Under the premise of a general difference between task-related and target-elicited saccades, we expected a similar shift in main sequence curves as in the experiment by Epelboim and colleagues (1997).

2.3 MATERIAL AND METHODS

Two experiments were run to compare classical target-elicited saccades (look condition) against task-related saccades, which were required for completing a discrimination task (discriminate condition). Experiment 1 was conducted to assess differences in saccade characteristics and differences in saccade RT distributions. Experiment 2 investigated changes in the saccade main sequence parameters following the presentation of targets at different eccentricities.

Participants

12 participants took part in experiment 1 (8 male, 4 female, ages 24–37) and another 12 participants took part in experiment 2 (7 female, 4 male, ages 21–31). All participants had normal or corrected to normal vision. In accordance with the World Medical Association's Declaration of Helsinki, written informed consent was obtained from all subjects prior to experimentation and the procedures of the experiment had been approved by the ethical committee of the University of Tübingen. Participants were paid 8 EUR per hour for taking part in the experiment.

Materials

In both experiments, participants sat in an adjustable chair in front of a CRT monitor (Sony GDM-FW 900, 100 Hz refresh rate, resolution 1600×1000) in a room with subdued light. A chin-rest provided support for the head at a viewing distance of 53 cm. An optical infrared head-mounted eye-tracking system was used to measure gaze at a sampling rate of 500 Hz (SR Research Eyelink II). A button box was used to collect manual responses. The eye-tracker and button box were connected to a dedicated computer which logged the data. Presentation of the experiment was controlled by custom-written software on a separate computer.

Stimuli

Two types of visual targets were designed for the two conditions (look, discriminate). In the look condition, the target consisted of a 3×3 pixel block (0.1 ° visual angle) with light gray color. In the discriminate condition, the same target was shown except that one pixel of the 3×3 pixels was missing (corresponds to a gap of ca. 0.04 ° visual angle, 2.4 minutes of arc). The gap was located either at the top or bottom of the target. In both conditions, a white border was drawn around the target to make it discernible in the visual periphery.

A uniform gray background with a luminance of $15 \, cd/m^2$ was shown throughout a trial. The target color was of a lighter gray with an average luminance of $22.5 \, \text{cd/m}^2$, which corresponds to a Weber contrast of 0.5 (contrast was calculated as $(I - I_b)/I_b$ where I represents the stimulus intensity and I_b the background intensity). The target's luminance contrast was adjusted separately for each participant to obtain a uniform degree of difficulty across participants. To do this, a block of trials of the discriminate condition was conducted at the beginning of the experiment. During this block of trials, the contrast was continuously adapted using the QUEST psychophysical procedure (Watson and Pelli, 1983), so that similar difficulty levels were obtained for each participant (on average 86%, SD 9.6%, correct responses).

Procedure

The basic experimental task required participants to make a saccade following target onset (Figure 8 A). Each trial commenced with the presentation of a fixation cross at the center of the computer screen. After a random delay with uniform distribution in the range of 0.5–1.5 s, a target appeared at $9\,°$ eccentricity either to the left or right of the central fixation cross. The central cross stayed visible throughout the trial and the target remained visible for 1.5 s. This was sufficient time for participants to make a saccade to the target.

In the discriminate condition, participants were instructed to identify the opening of the target (top or bottom). Due to the small size of the gap, a saccade to the target was necessary in order to achieve this. After the target disappeared, participants indicated whether the target's opening was at the top or bottom by pressing the corresponding button on the button box (up or down). In this condition, participants were told to identify the target as accurately as possible. No specific instruction was given with respect to saccade or response speed. In the look condition, participants were instructed to look at the target as quickly as possible. After the target disappeared, participants responded by pressing the up button on the button box to confirm trial completion and to keep the sequence of events consistent with the discriminate condition.

A similar procedure was used by two previous studies (Trottier and Pratt, 2005; Guyader et al., 2010). However, it is not clear which role time pressure played in these experiments. In other experiments, time pressure was induced by an instructional emphasis on response speed (Reddi and Carpenter, 2000) or limitation of target presentation time (Montagnini and Chelazzi, 2005). The former was also true for the study by Trottier et al. (2005) and the latter applied to the work by Guyader et al. (2010) (target presentation time 500 ms). We addressed these issues in our own experimental design. First, target presentation time was long enough (1.5 s) for participants to perform a saccade and still have sufficient time to discriminate the target. Second, target presentation and response input was separated into two phases of the trial. Participants first looked at the target. After that, the target disappeared and a question mark symbol prompted participants to press the appropriate response button. Early termination of a trial by participants through a premature response was therefore not possible.

Most everyday tasks consist of simple goal-directed behaviors (Land et al., 1999; Land and Hayhoe, 2001; Dickinson and Balleine, 1994). Feedback on the results of an action is usually available in such behaviors (Salmoni et al., 1984). For example, participants clearly perceived whether they successfully tapped on the targets in the pointing task presented by Epelboim et al. (1997). To provide clear feedback in the purely visual task that was used in the current experiment, a pictogram, which was either a circle for correct or cross for incorrect actions, was shown after each trial. In the discriminate condition, feedback was contingent on a participant's response and the actual location of the opening. In the look condition, positive feedback was presented if a saccade to the target and the confirmatory button-press occurred within the respective time windows.

In total, each participant performed 480 trials in experiment 1. These comprised six blocks of 40 trials for each condition. The eye-tracking system was re-calibrated after each block. Regular 5 minute breaks were provided in intervals of three blocks of trials, during which the eye-tracker was removed. The order of conditions was fully counter-balanced between participants, half of which began a session with the discriminate or look condition. The entire experimental session lasted about 120 minutes.

The same procedure was also used in experiment 2 with the modification that targets were presented randomly at different eccentricities in the range of 1.5° to 20°. Since observers' head motion was constrained by the eye-tracking equipment, we chose eccentricities close to those observed in natural gaze behavior (Sanders, 1970; Stark et al., 1975). As a result, the eccentricities of these locations were smaller than those used by previous authors (using a head-free tracking system, Epelboim et al., 1997 presented eccentricities up to 45°).

Four of the participants were tested in sessions which were held on successive days. Each experimental session lasted ca. 90–120 minutes. 900 data points were collected for these four participants. The remaining 8 participants were tested in single sessions lasting 120–160 minutes. During these sessions, conditions were presented randomly in blocks of 30 trials. This was done to minimize potential effects of day-to-day variability in performance due to different levels of fatigue. 360 data points were collected per participant during these recording sessions.

Data Analysis

Saccade detection was carried out by the Eyelink II system using a velocity (22°/s) and acceleration threshold (3800°/s^2). The primary measures used to characterize saccadic eye movements were saccade reaction time (RT), peak velocity, duration, and gain. Saccade RT was defined as the time between the onset of the target and initiation of the movement. Saccade gain was defined as the size of the saccade divided by the step size, i.e., the distance between the location of gaze before the saccade and the target.

Data from the following trials were removed prior to the analysis: Trials with blinks during the critical time period shortly before or after the target onset, missed trials (no saccade or RT greater than 700 ms), anticipatory saccades (RT smaller than 50 ms), and inaccurate saccades with gains larger than 1.5 or smaller than 0.5.

In total 5760 data points were collected during experiment 1. 190 data points (3%) were removed due to application of the outlier rules. Analyses were carried out on the remaining data points. For saccade RT data, per-participant and condition cutoffs were employed (Ratcliff, 1993). These cutoffs removed data points > 1.5 SD (median amount of points removed 7.5%, max. 13%). 6480 datapoints were collected in experiment 2. Of these, 460 data points (7%) were removed due to the outlier criterions.

If not indicated otherwise, paired two-tailed t-tests were employed for the comparison of mean differences ($\alpha = 0.05$) and mean-centering was performed for the computation of confidence intervals (Baguley, 2012). Becker's g, which is also known as Glass's Δ was used as a measure of effect size (Kline, 2005). This is the mean difference between conditions divided by the baseline standard deviation (i.e., the SD of the look condition).

2.4 RESULTS

Experiment 1: Looking vs. Discriminating

Saccade reaction time, peak velocity, duration, and gain were measured across two conditions of a saccade task. In condition 1 (discriminate condition) participants made a saccade to a target in order to identify it (see materials and methods section and Fig. 8). The target was a Landolt-square optotype, i.e., a small square with an opening on either the top or bottom (similar to Yeshurun and Carrasco, 1999). Condition 2 (look condition) was identical to the discriminate condition, except that the square was shown without an opening. In this condition, participants were instructed to fixate the square as quickly as possible. Assuming that task-related saccades are categorically different from classical target-elicited saccades and that the results of previous experiments (e.g., Epelboim et al., 1997) can in part be explained by this difference, we expected shorter saccade reaction times and higher peak velocities in the discriminate condition.

Saccade RT and velocity were compared across the two conditions to assess whether saccades were faster and started earlier in the discriminate condition. Mean saccade reaction time in the look condition was 194 ms (SD 40 ms) compared to 163 ms (SD 32 ms) in the discriminate condition (individual means are shown in Fig. 8 B). This difference (95% confidence interval of difference: 15–52 ms, effect size $\Delta = 0.90$) is statistically significant ($t(11) = 3.4$, $p < 0.01$).

Mean saccade peak velocity was 382 °/s (SD 41) in the look condition and 393 °/s (SD 41) in the discriminate condition. This difference

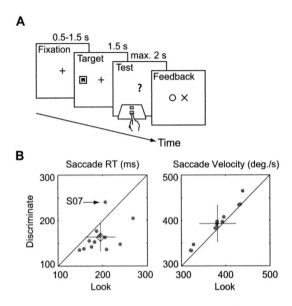

Figure 8: Experimental task and results (exp. 1). **A** Schematic of the discriminate task. Participants fixated a central cross. This was followed by target onset either to the left or right of the fixation cross. Then, participants looked at the target and identified the location of the gap in the square. After this, the target disappeared and participants responded with the appropriate button press on a button box. Feedback was then presented depending on the response and actual gap location. The sequence of events was similar in the look condition except that no discrimination had to be carried out and participants were instructed to look at the target as quickly as possible. Here, participants confirmed trial completion by pressing the up button on the button box. Positive feedback was presented if a correct saccade was performed and the button response was given within the time window. **B** Scatterplots of saccade properties with participant means, standard deviation (cross), and 95% confidence intervals (diamond) show shorter RTs and faster velocities in the discriminate condition. Data from participant S07 exhibits a potentially abnormal RT distribution (see text).

(95% confidence interval of difference: 8–17 °/s, $\Delta = 0.28$) is statistically significant ($t(11) = 5.5, p < 0.01$).

A small but significant difference in saccade duration was found between both conditions. Mean saccade duration was 48.6 ms (SD 2.9 ms) in the look condition and 47.6 ms (SD 3.0 ms) in the discriminate condition. This difference (95% confidence interval of difference: 0.3–1.7 ms, $\Delta = 0.32$) is statistically significant ($t(11) = 2.5, p < 0.05$).

Saccade velocity and duration are known to be functions of saccade amplitude. To test whether the increase in peak saccade velocity could be the result of different amplitudes, saccade gain was compared. In both conditions, saccade gain was close to one. In the look condition gain was 1.015 (SD 0.03) and 1.022 (SD 0.03) in the discriminate condition. This difference is not statistically significant ($t(11) = 1.69$, $p = 0.12$).

To assess changes in saccade characteristics over the course of the experiment, best linear fits were obtained across trials (Fig. 9). This showed a positive correlation of saccade RT in the discriminate condition ($p < 0.01$, $R^2 = 0.45$) and a negative correlation of peak saccade velocity in the look condition ($p < 0.01$, $R^2 = 0.26$), which could indicate that the difference in RT between both conditions decreased over the experiment while it increased for peak velocity.

A more thorough analysis of RT data was conducted to explain observable differences in RT distributions from raw RT histograms (Fig. 10 B). Sequential-sampling models such as the LATER model have been used in previous studies to successfully explain the shape of saccade RT distributions (Carpenter, 1988; Carpenter and Williams, 1995). The LATER model assumes that saccade initiation is determined by the accumulation of sensory evidence over time (Fig. 10 A). Specifically, it considers two main variables: a) the rate of rise of the decision signal (Reddi et al., 2003) and b) the decision threshold (Carpenter and Williams, 1995; Reddi and Carpenter, 2000).

Maximum likelihood estimates of these variables were obtained on individual data and separately for each condition from the main part of RT distributions. Bimodal RT distributions were visible in the data of two participants, with the first mode around 100 ms, which is typically associated with express saccades (Fischer and Boch, 1983; Fischer et al., 1993). In line with previous research (e.g., Montagnini and Chelazzi, 2005) parameters were fitted to the non-express part

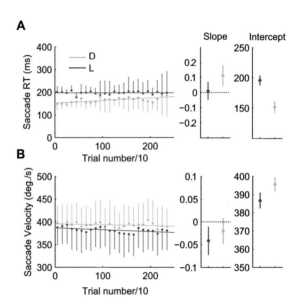

Figure 9: Changes in saccade parameters over time (exp. 1). Best linear fits across mean data. The data was binned in blocks of 10 trials. Data points show mean and variance for saccades performed in the look (L) and discriminate (D) condition. These trends suggest that the differences in saccade RT decreased over time while the difference in velocity increased.

of the distribution in these datasets. Kolmogorov-Smirnov tests were carried out on each dataset to verify that reciprocal RT data was compatible with the assumption of a normal distribution, as predicted by the model. This was the case for all datasets ($p > 0.1$) except for the data from one participant (participant S07, $p < 0.01$). Inspection of this data showed an extreme spread of RTs in both conditions, which could be evidence for fatigue. The data of this participant was therefore excluded from further analyses (this was also the only dataset that exhibited longer RTs in the discriminate condition, see Fig. 8 B).

Average predicted distributions and parameter values are shown in Fig. 10 C. The theoretical distribution during the discriminate condition is characterized by a negative shift of the mode and decreased variability, which is evident from the shorter tail. Comparison of model parameters showed a significantly higher rate ($t(10) = 3.2, p < 0.01$, 95% confidence interval of difference: 0.2–1, $\Delta = 0.53$) and only a small difference in threshold, which is not statistically significant ($t(10) = 0.8, p = 0.4$). This suggests that the primary difference of RT data between both conditions was due to a change of the rate of rise of the decision signal, similar to previous findings which related changes in RT to a change in the rate of information supply (Reddi et al., 2003) or effects of perceptual urgency (Montagnini and Chelazzi, 2005).

Overall, the results clearly illustrate a fundamental difference between target-elicited and task-related saccades. In line with our hypothesis, task-related saccades exhibited shorter RTs and higher peak velocities. These findings are similar to those previously attributed to the effects of motor coordination (Epelboim et al., 1997; Snyder et al., 2002). In addition, a comparison of saccade RT distributions using LATER model fits shows differences in the rate parameter – a finding which was previously attributed to effects of perceptual urgency (Montagnini and Chelazzi, 2005).

Experiment 2: Saccade Main Sequence

Saccade velocity and duration is strongly related to the amplitude of the required saccade. This relationship has been referred to as the saccade *main sequence* (Bahill et al., 1975; Leigh and Zee, 2006). Existing models explain this dependency as a result of duration-accuracy

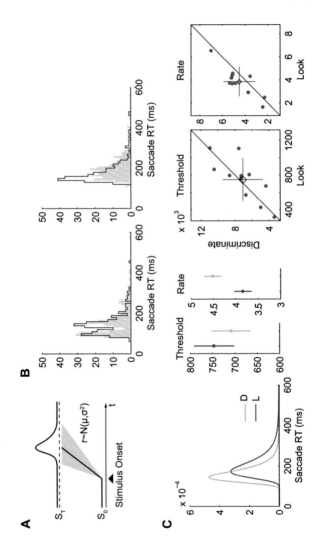

optimizations, which lead to optimal trajectories for any given target eccentricity (Harris and Wolpert, 2006; Goossens and Van Opstal, 2006). In addition, previous work suggests that saccade kinematics are also influenced by a variety of other aspects, for instance, the need to carry out an arm movement in coordination with an eye movement (Epelboim et al., 1997; Snyder et al., 2002).

Such modulations may not necessarily be the result of coordinated motor actions. The results from our first experiment suggest that task-related saccades in general, even in the absence of oculomanual actions, might have higher peak velocity than target-elicited saccades. In experiment 2, we extend this finding by examining saccade velocities across a range of amplitudes. With regard to the results of previous studies on motor coordination (Epelboim et al., 1997; Snyder et al., 2002), we expected main sequences of task-related saccades to show different properties (e.g., a steeper rise in velocity or higher saturation velocity) than target-elicited saccades.

Figure 10 *(preceding page)*: RT model, observed and theoretical RT distributions (exp. 1). **A** Schematic of the LATER model. The model assumes that saccades are initiated once a decision signal rises from its baseline level S_0 to a threshold S_T after target onset. The rate of rise r exhibits trial-to-trial variability, which is modeled by a normal distribution. The distribution of RTs resulting from this process is shown above. **B** Observed RT distributions for two participants. Filled histograms show data for the look condition, outlines show data for the discriminate condition. Left: One of the observed bimodal distributions. For these, distribution parameters were estimated from the non-express part of the distribution (right mode). Right: Example for a more commonly observed unimodal distribution. **C** Theoretical RT distribution as predicted by the LATER model for RT data in the look (L) and discriminate (D) condition. Right: Model parameters (threshold and rate) with 95% confidence intervals and scatterplots of the parameter distributions with mean, standard deviation (cross), and 95% confidence intervals (diamond) showing that the likely explanation for differences in the distributions is a change in the rate of rise.

To analyze changes in peak velocity across amplitudes, an exponential main sequence function of the form $V = V_{max}(1 - e^{-A/C})$ was fitted to individual peak velocity data (Epelboim et al., 1997; Leigh and Zee, 2006). Here, V_{max} denotes the saturation velocity and A the saccade amplitude. The time constant C represents the amplitude at which 63% of the saturation velocity is reached and thus describes how quickly saturation is attained. Posterior amplitudes were used for fitting, i.e., the amplitudes that were actually performed, which were sometimes slightly longer or shorter than the required amplitudes.

Fig. 11 A shows a typical distribution of peak velocity data points and the resulting fit of the theoretical model (black line). Fig. 11 C shows the theoretical main sequences and parameters for both conditions following parameter averaging. On average, saccade duration was predicted best by $523\,(1 - e^{-A/6.8})$ in the discriminate condition and by $496\,(1 - e^{-A/6.9})$ in the look condition. A statistical comparison of model parameters shows a significant difference in the saturation velocity V_{max} $(t(11) = 5.3, p < 0.01, 95\%$ confidence interval of difference: 18–38, $\Delta = 0.38)$ but not in the time constant C $(t(11) = 0.45, p = 0.66)$.

A linear relationship between saccade duration and amplitude was assumed for saccades larger than four degrees (Carpenter, 1988; Garbutt et al., 2001). On average, saccade duration was predicted best by $2.18\,A + 31.9$ in the discriminate condition and by $2.33\,A + 31.5$ in the look condition. A comparison of parameter averages shows a significant difference in the slope parameter $(t(11) = 2.6, p < 0.05,$ 95% confidence interval of difference: 0.03–0.24, $\Delta = 0.46)$ and an insignificant difference in the intercept parameter $(t(11) = 0.6, p = 0.59)$.

An additional ad hoc analysis was performed for the data of participant S3, which showed a distinctive scatter of data points below the main sequence curve in the look condition. This resulted in a large difference in the time constant parameter (Fig. 11 B). Scatter below the main sequence curve is known to indicate fatigue (Schmidt et al., 1979). To analyze this, we identified all data points outside a 95% prediction interval around the obtained main sequence. Further separation according to trial number showed that the majority of these outliers (34 of 37 points, > 90%) occurred in the second half of the experimental session ($\chi^2 = 25, p < 0.01$). This suggests that, at least for this participant, fatigue due to repetitions might be one important factor which could explain the main sequence parameter differences, specifically, the difference in the saturation time constant. Overall, however, the results indicate that task-related saccades exhibit an increase in the saturation velocity of the velocity main sequence and a decrease in the slope of the duration main sequence in comparison to target-elicited saccades.

2.5 DISCUSSION

The present study compared the characteristics of *task-related saccades*, which supported a visual discrimination task, and classical *target-elicited saccades*, which were not followed by such a task. Experiment 1 showed that task-related saccades exhibit shorter reaction times, higher peak velocities, and shorter durations than target-elicited saccades. This is even more surprising since participants were instructed to perform target-elicited saccades as quickly as possible whereas emphasis was put on (task) accuracy when performing task-related saccades. The LATER sequential-sampling model (Carpenter and Williams, 1995) was used to model saccade RT distributions. An analysis of model fits revealed that differences between RT distributions of both saccade types could be explained by assuming a steeper rate of rise in the decision signal. Experiment 2 tested how the need to perform a discrimination task at the saccade endpoint affected the saccade main sequence, the relationship between saccade peak velocity, duration, and amplitude. Our results show an increase in the saturation velocity of the velocity main sequence and a decrease in the slope of the duration main sequence for task-related saccades.

Three basic explanations for the general differences in saccade RT and velocity can be excluded. First, it is well known that fundamental stimulus properties (e.g., luminance contrast) exert an influence on behavioral response characteristics and could have generated faster responses in one condition (Cattell, 1886; Mansfield, 1973; Reulen, 1984; Reuter-Lorenz et al., 1991). Considering the small differences between the two targets, this explanation is unlikely. Second, the change in peak velocity could have been a concomitant of increased saccade gain. We dismiss this explanation by noting that the measured

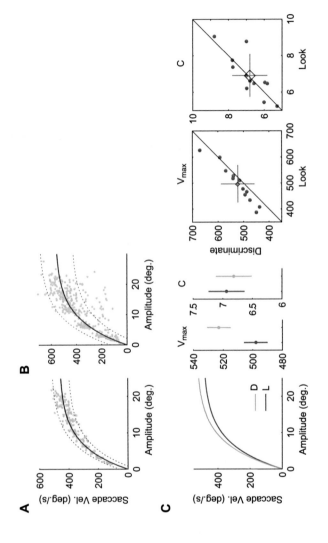

48

differences in gain were very small and not statistically significant. Third, an explanation in terms of dual-task effects on saccade RT, which were previously reported in saccade and discrimination tasks (Kowler et al., 1995; Hoffman and Subramaniam, 1995; Deubel and Schneider, 1996), is not applicable, since the location of the saccade target and discrimination target was not dissociated experimentally.

Previous studies which examined the functional variability of saccade properties obtained similar results, for example, higher saccade velocities and shorter reaction times during oculomanual actions such as pointing or grasping (Epelboim et al., 1997; Snyder et al., 2002) or object identification under time pressure (Montagnini and Chelazzi, 2005). How do these results relate to our findings and how can our findings be explained without invoking mechanisms of motor coordination or time pressure?

We speculate that differences between task-related and target-elicited saccades could be related to repetitions and motivation. Previous work has shown that massed repetitions of target-elicited saccades can result in a decrease in peak velocity (Schmidt et al., 1979; Fuchs and Binder, 1983; Straube et al., 1997; Chen-Harris et al., 2008; Prsa et al., 2010). One explanation for our findings could therefore be that task-related and target-elicited saccades are affected differently by

---

Figure 11 *(preceding page)*: Saccade velocity main sequence (exp. 2). **A** Example for a commonly observed distribution of saccade velocities as a function of amplitude. The figure shows the data for one participant in the look condition. The solid line shows the best fit of $V = V_{\max}(1 - e^{-A/C})$. Dotted lines show 95% prediction intervals. **B** The data from one participant in the look condition shows a significant number of datapoints below the main sequence curve (outside the prediction interval). This could indicate fatigue. **C** Left: Theoretical main sequence curves for average parameters in the look (L) and discriminate (D) condition. Right: Mean model parameters with 95% confidence intervals and scatterplots of the parameter distributions with mean, standard deviation (cross), and 95% confidence intervals (diamond). This shows a significant difference in saturation velocity $V_{\max}$.

repetitions. Indeed, our results show a differential effect of repetitions on saccade characteristics, with peak velocity decreasing slightly in the look condition and saccade RT increasing in the discriminate condition (exp. 1). The work by Prsa et al. (2010) shows that repeated saccades are not affected by muscular fatigue but by higher-order *mental fatigue*. In this respect, the current results could be explained by the interaction of two effects. First, a general arousal-related effect due to the monotonous nature of the task. Second, a more specific effect related to *motor readiness* due to motivational differences between the two saccade types (De Brabander et al., 2002). A decline in arousal could have caused the general decrease in saccade velocity over time (exp. 1). In addition, presentation of negative feedback increases arousal (De Brabander et al., 2002). This could explain why the decline in saccade velocity was less pronounced in the discriminate condition. Due to the presence of the discrimination task, more errors, and thus more negative feedback was presented in the discriminate condition compared to the look condition.

Previous writers have suggested that motivation influences saccade characteristics (e.g., Chen-Harris et al., 2008). Evidence has been provided, showing that saccade characteristics can be shaped by rewarding saccades (Kawagoe et al., 1998; Lauwereyns et al., 2002; Takikawa et al., 2002; Milstein and Dorris, 2007; Stritzke et al., 2009; Shadmehr et al., 2010; Navalpakkam et al., 2010; Madelain et al., 2011b,a). In this regard, target-related saccades could be inherently more rewarding than classical target-elicited saccades. This could be the case because task-related saccades support completion of a meaningful task, which addresses competency-related needs (White, 1959; Deci and Ryan, 2000). Following the argumentation of Chen-Harris et al. (2008), this inherent reward value could decline with repeated stimulus presentations. This could explain the measured increase in RT over time in the current experiment, which was pronounced in the discrimination task. This explanation assumes that saccade characteristics can be affected by explicitly rewarding saccades, as well as by the reward value that is inherently associated with the task supported by the saccade. To the best of our knowledge, the existence of such an indirect influence is yet to be demonstrated and merits future investigation.

Differential motivational levels could also explain the obtained change in RT distributions which was revealed by the analysis using the LATER model (Carpenter and Williams, 1995). This model predicts saccade RT on the grounds of a rising decision signal with a variable rate of rise and decision threshold (Fig. 10 A). A functional interpretation of the LATER model relates this decision signal to the accumulation of sensory evidence about the correct saccade choice. Evidence for this is provided by previous research, which found that manipulations of prior target probability and time pressure affect the baseline level or threshold of the hypothesized signal (Carpenter and Williams, 1995; Reddi and Carpenter, 2000). Changes of the rate of rise of the decision signal were associated with the available amount of sensory information relevant for the decision, for instance, the coherence of dot movements in a random-dot kinematogram (Reddi et al., 2003).

Best fits of the LATER model to the current data revealed that task-related saccade distributions exhibit a steeper rate of rise than target-elicited saccades. A change in the rate of rise was also observed by Montagnini and Chelazzi (2005) in their comparison of saccade RTs to visual targets and saccades that were followed by a visual discrimination task under time pressure. This is incompatible with previous work by Reddi and Carpenter (2000), which predicts that time pressure should lead to a change in the threshold parameter. Furthermore, Montagnini and colleagues showed that a gradual increase in time pressure did not result in a gradual decrease in saccade RT (see Montagnini and Chelazzi, 2005, experiment 2). Together, this suggests that the results of Montagnini and Chelazzi may not primarily reflect the time pressure that was associated with the discrimination task but, similar to the results of our own study, a more general influence of the visual task which followed the saccades.

Previous studies related a change in the rate of rise of the decision signal to the rate at which information is supplied to the saccadic choice process (Reddi et al., 2003). Neither the results of Montagnini and Chelazzi (2005) nor our own results can be explained on the grounds of an unbalanced supply of information since target onset was equally perceptible in both conditions. However, a possible explanation for the change in rate of rise in line with this interpretation of LATER's parameters could be that participants were less *efficient* in using the available information in target-elicited saccades, as a result of motivational differences. For instance, parietal and frontal brain areas (e.g., lateral intraparietal area or frontal eye fields), which are known to be implicated in saccade generation and are likely implementations of an internal decision mechanism, are also known to be affected by the magnitude of expected rewards (Platt and Glimcher, 1999). This could be a partial explanation for the data observed by Montagnini and Chelazzi (2005), instead or in addition to the assumed effect of perceptual urgency.

A comparison of task-related and target-elicited velocity main sequences shows a higher maximal velocity (saturation velocity) and a small difference in the saturation time constant. This observation is quite similar to that of Epelboim et al. (1997), who observed higher saturation velocities when participants tapped rather than looked at targets in succession. Differences in main sequence curves due to fatigue can be expected to lead to slower saturation. This is exemplified by the data of one participant in our experiments (Fig. 11 B). These data show a large saturation time constant difference between the two conditions, primarily due to a distinctive scatter of data points below the main sequence curve in the look condition. This scatter is similar to the observations by Schmidt et al. (1979) who measured a fatigued observer. We did not observe similar patterns in the other data sets nor a significant overall difference in the saturation time constant parameter in our data. It is therefore unlikely that the overall difference in saccade velocities reflects a difference in the level of fatigue. Instead, following our previous argument, the increase in saccade velocity which is evident from the comparison of saturation velocity parameters could reflect the increased strength of the saccade target signal (see also Snyder et al., 2002). This signal could primarily be influenced by salience and motivation, rather than effects of oculomanual coordination as Epelboim and colleagues (1997) assumed.

In conclusion, the present study highlights a fundamental difference between task-related and classical target-elicited saccades. Task-related saccades exhibit shorter reaction times and higher peak velocities. These differences are also evident in systematic changes in saccade RT distributions and the relationship between saccade velocity, duration, and amplitude (main sequence). The present experiments also show that previous task-specific explanations of differences between task-related and target-elicited saccades might be too narrow in scope. Further experimentation is required to test alternative explanations, for instance, ideas put forth by neurophysiological research, which indicates a modulation of saccade characteristics by motivational aspects of the task.

ACKNOWLEDGEMENTS

This research was supported by a PhD stipend from the Max Planck Society (HJB), the Information Technology Program through the state of Baden-Württemberg, Germany (BW-FIT) (HJB, LLC), and by the WCU (World Class University) program funded by the Ministry of Education, Science and Technology through the National Research Foundation of Korea (R31-10008) (HHB).

COPYRIGHT

This is an open-access article distributed under the terms of the Creative Commons Attribution License, which permits unrestricted use, distribution, and reproduction in any medium, provided the original author and source are credited.

3

EYE MOVEMENTS DURING PURSUIT TRACKING

This chapter has been reproduced from an article published in Experimental Brain Research: Bieg, H.-J., Bresciani, J.-P., Bülthoff, H. H., & Chuang, L. L. (2013). Saccade reaction time asymmetries during task-switching in pursuit tracking. *Experimental Brain Research*, 230(3), pp. 271-281.

3.1 ABSTRACT

We investigate how smooth pursuit eye movements affect the latencies of task-switching saccades. Participants had to alternate their foveal vision between a continuous pursuit task in the display center and a discrete object discrimination task in the periphery. The pursuit task was either carried out by following the target with the eyes only (ocular) or by steering an on-screen cursor with a joystick (oculomanual). We measured participants' saccadic reaction times (SRTs) when foveal vision was shifted from the pursuit task to the discrimination task and back to the pursuit task. Our results show asymmetries in SRTs depending on the movement direction of the pursuit target: SRTs were generally shorter in the direction of pursuit. Specifically, SRTs from the pursuit target were shorter when the discrimination object appeared in the motion direction. SRTs to pursuit were shorter when the pursuit target moved away from the current fixation location. This result was independent of the type of smooth pursuit behavior that was performed by participants (ocular/oculomanual). The effects are discussed in regard to asymmetries in attention and processes that suppress saccades at the onset of pursuit.

3.2 INTRODUCTION

Saccades are discrete shifts of the eyes that place the image of an object of interest on the fovea for detailed inspection. Smooth pursuit are a different class of eye movements. They are much slower than saccades and move the eyes in a continuous fashion when following a moving stimulus (Orban de Xivry and Lefèvre, 2007). The coordination of saccades and smooth pursuit eye movements is not well understood. Both types of eye movements are typically examined during fixation of a moving object. Here, smooth pursuit eye movements are complemented by saccades that automatically "catch-up" with the moving object once fixation error accumulates (e.g., de Brouwer et al., 2002b). But this is not the only class of joint ocular behavior in everyday tasks. Alternating smooth pursuit and saccadic eye movements also occur when the observer switches between tasks. For instance, drivers move their gaze back and forth between following the road and the dashboard instruments. There, saccades may also occur before or after smooth pursuit. This interaction between both classes of eye movements during task-switching has received considerably less attention.

Studies that have investigated saccade initiation during pursuit provide some insight in this regard (Tanaka et al., 1998; Kanai et al., 2003; Khan et al., 2010; Seya and Mori, 2012). For example, in the study by Tanaka et al. (1998), participants pursued a moving stimulus and were instructed to switch their gaze to a second stimulus after its onset. The results show that saccade reaction times (SRTs) are asymmetric: Saccades to targets in the motion direction of pursuit exhibit shorter SRTs than saccades against the motion direction. Together with the finding that covert attention facilitates the detection of and the response to peripheral stimuli (Posner, 1980, see also Deubel and Schneider, 1996; Kowler et al., 1995; Hoffman and Subramaniam, 1995), this phenomenon has been linked to shifts of spatial attention in the direction of pursuit in anticipation of the pursuit target's future position (van Donkelaar and Drew, 2002; Khan et al., 2010; Seya and Mori, 2012; but see also Heinen et al., 2011; Lovejoy et al., 2009; Prinzmetal et al., 2005).

However, anticipatory behavior during smooth pursuit has also generally been observed in basic ocular pursuit tasks (Shagass et al., 1976; Mather and Putchat, 1983; Gauthier et al., 1988; Van Gelder et al., 1990; Koken and Erkelens, 1992; Sweeney et al., 1994; Van Gelder et al., 1995b,a; Kathmann et al., 1999). For example, Van Gelder et al. (1990) compared pursuit performance in a standard ocular pursuit and a more naturalistic visual analysis condition. The results show a larger fixation error in the pursuit-only condition due to an increased number of anticipatory saccades that interrupted smooth pursuit. Similar results were obtained by Koken and Erkelens (1992), who showed that smooth pursuit was less frequently interrupted by saccades when it was performed during a manual tracking task rather than a basic ocular pursuit task.

The goals of the current study are twofold. First, we replicate the finding by Tanaka et al. (1998) and others (see above), namely SRT asymmetries *from* pursuit. We do this to test whether these asymmetries hinge on anticipatory behaviors that commonly occur in basic, laboratory ocular pursuit tasks (Van Gelder et al., 1995a). Second, we extend the analysis of SRT asymmetries to a more comprehensive task-switching scenario. Previous studies limited their investigation to saccades *from* pursuit and did not consider saccades *back* to the pursuit target. Such back-and-forth motion of gaze is common when a task is switched and later resumed, for instance, in our earlier example of driving. An experiment that required such task switching was recently conducted by Jonikaitis et al. (2009). Nonetheless, this study was not primarily designed to address our current question. Thus, it did not investigate motion-related differences in saccade onsets.

To address these two questions, we used an experimental paradigm in which foveal vision was shared between a continuous pursuit task and a secondary, discrete object discrimination task. We measured the SRTs of saccades that moved the eye *away* from the pursuit target to the discrimination object and *back* to the moving pursuit target. Two variants of this task were presented. Participants followed the target either with their eyes only, which replicates the basic ocular pursuit condition that has been used by previous studies, or by steering a cursor with a joystick. In the latter condition, smooth pursuit provides important task-related information. This is expected to result in more natural eye movements, in particular, a reduction of anticipation as it can be observed in basic ocular pursuit tasks (Van Gelder et al., 1995a; Koken and Erkelens, 1992).

Participants

Twelve participants took part in the experiment (9 male, 3 female, age: 21–36 years). All participants had normal or corrected to normal vision. A vision test was conducted to verify this prior to the experiment (FrACT test Bach, 2007, logMAR < 0 for all participants). In accordance with the World Medical Association's Declaration of Helsinki, written informed consent was obtained from all subjects prior to experimentation and the procedures of the experiment had been approved by the ethical committee of the University of Tübingen. Participants were paid 8 EUR per hour for taking part in the experiment.

Materials

Participants sat in an adjustable chair in front of a TFT monitor (Samsung 2233RZ, 120 Hz refresh rate, resolution 1680 × 1050, see also Wang and Nikolić, 2011). A chin-rest provided support for the head at a viewing distance of 57 cm. An optical infrared head-mounted eye-tracking system was used to measure gaze at a sampling rate of 500 Hz (SR Research Eyelink II). A potentiometer joystick (0.18° angular accuracy, sampling rate 120 Hz) was mounted under the table within comfortable reach for the participants. The joystick was moved to the side of the dominant hand for each participant. With the other hand, participants pressed the cursor keys on a keyboard.

Stimuli

Two types of stimuli were used in the experiment. Stimuli for the pursuit task consisted of differently colored vertical bars. The pursuit target was blue (RGB 180, 180, 255) and subtended 1.2° (visual angle), the cursor was orange (RGB 255, 255, 100) and subtended 0.9°. A second stimulus type was used for the object discrimination task. This stimulus consisted of a small square (0.2°) of white color (RGB 200, 200, 200). A small gap was present at one of the four sides of the square (size 0.03°, 1.8 minutes of arc). A white border was drawn around the target to make it discernible in the visual periphery. All stimuli were presented against a uniform gray background (RGB 100, 100, 100).

Task

The primary *pursuit tracking* task required participants to steer an on-screen cursor using a joystick (see Fig. 12 A, Fig. 13). By moving the joystick to the left or right, participants controlled the horizontal velocity of the cursor. The instruction was to move the cursor "as close as possible" to a computer-controlled pursuit target. The pursuit target moved horizontally in a sinusoidal path around the center of the computer screen with an amplitude of 4.3° and frequency of 0.25 Hz. This task was performed continuously in blocks, each block lasting 128 s.

The secondary *object discrimination* task required participants to look at and identify a discrimination object. Participants were instructed to discern the side of the target where the gap was located (top, bottom, left, right). Due to the small size of the gap, a saccade to the target was necessary in order to achieve this. After participants looked at the target to determine the gap they responded with one of the four corresponding arrow keys on the keyboard.

Each pursuit block was subdivided into trial epochs of 8 s. This subdivision was not made explicitly apparent to the participants and served as a framework to control the onset of the discrimination stimulus in relation to the pursuit target motion (400 ms before the pursuit target crossed the middle of the screen). To make the repetitive appearance of the discrimination stimulus less predictable, the onset timing and location of the stimulus was varied. The stimulus appeared either 1.3 or 3.6 s after the start of an epoch at an eccentricity of 13°. The discrimination task was scheduled such that discrimination objects appeared with equal probability and frequency either in the same direction as the movement (to condition) of the pursuit target or in the opposite direction (away condition, see Fig. 12 B).

Two conditions of the pursuit task were presented. In the oculomanual condition, participants controlled an on-screen cursor as

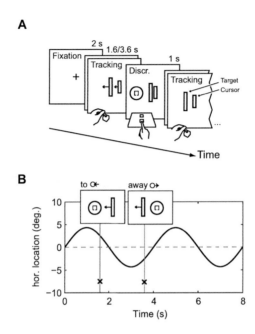

Figure 12: **A** Schematic of the experimental task. Each block started with an auditory warning signal and a fixation cross. This was followed by a continuous tracking block lasting 128 s. Here, participants controlled the horizontal speed of an on-screen cursor by moving the joystick to the left or right. They were instructed to follow the pursuit target as closely as possible, which moved horizontally on a sinusoidal path. The pursuit task was interrupted by a secondary task. This was an object discrimination task in which participants had to recognize the opening of a square optotype. **B** The pursuit target performed two full cycles every 8 seconds (one epoch). During each epoch the discrimination object was presented randomly either 1.6 or 3.6 s into the epoch on the left or right side of the screen. The time and location defined whether the discrimination object was presented while the pursuit target was moving *to* the location of the discrimination object or while it was moving *away*.

described before. In the ocular condition, no cursor was present and participants were simply instructed to look at and follow the motion of the pursuit target.

Design and Procedure

The experiment was planned as a within-subjects factorial design with the factors task (ocular, oculomotor) and motion direction relative to the location of the discrimination object (to, away). The presentation of the task factor was balanced between subjects, motion direction was varied randomly.

During a session, tasks were presented in several runs. Each run took ca. 15 minutes including set-up and calibration of the eye-tracker. During a run, participants performed 5 blocks of the experimental task. Regular 10 minute breaks were provided after each run, during which the eye-tracker was removed. The order of task conditions was fully counter-balanced between participants, half of which began a session with ocular pursuit or oculomanual pursuit. The entire session lasted ca. 120 minutes.

Data Analysis

Saccade detection was carried out by the Eyelink II system using a velocity ($22\,°/s$) and acceleration threshold ($3800\,°/s^2$). The primary measures used to characterize saccadic eye movements were saccade reaction time (SRT), saccade amplitude, and gain. SRT was defined as the time between the onset of the discrimination object and initiation of the movement. SRTs for saccades back to the pursuit target (inward) were measured from fixation onset on the discrimination object to the beginning of a return saccade to the pursuit target. Saccade gain was defined as the size of the saccade divided by the step size, i.e., the distance between the location of gaze before the saccade and the target. For inward saccades, the pursuit target's location at the saccade onset was used to calculate gain.

Data from the following trials were removed prior to saccade analysis: Trials with blinks during the critical time period shortly before or after the target onset, missed trials (no saccade or RT greater than 800 ms), anticipatory saccades (RT smaller than 50 ms), inaccurate saccades with errors larger 2° visual angle, and trials with blinks shortly before or after the inward saccade. Based on this method, 102 data points of 1998 were removed (5.1%). The median number of data points remaining per participant and condition was 39 (min. 30).

For the frequency domain analysis, Fourier transforms were computed from whole 128 s blocks. Periods during which the eye moved to the discrimination object were removed by linearly interpolating between the eye position shortly before the outward and shortly after the inward saccade. The phase shift between signals was computed by subtracting the phase of the eye from the phase of the target signal at the fundamental frequency of 0.25 Hz (see also Vercher and Gauthier, 1992).

If not indicated otherwise, data plots show Cousineau-Morey confidence intervals (see Baguley, 2012; Morey, 2008).

3.4 RESULTS

Separate repeated-measures ANOVAs were employed to analyze outward and inward saccades. The primary dependent variable was saccade reaction time (SRT). In addition, saccade amplitude, gain, and endpoint error were computed to test whether differences in SRTs could be attributed to differences in the saccade magnitude or accuracy. The primary factors under investigation were the pursuit target movement direction (to or away from the discrimination object) and the type of tracking (ocular and oculomanual). The onset time of the discrimination object during the tracking epoch was treated as a third factor since the predictability of target onsets by the observers potentially differed between both onset times (onsets occured either early during the epoch at 1.3 s or late during the epoch at 3.6 s).

Outward Saccades

On average, saccades to the discrimination object (outward) were initiated after 232 ms. In both pursuit conditions, saccades that were initiated while the pursuit target moved to the discrimination object exhibited shorter RTs (222 ms) compared to saccades that started when

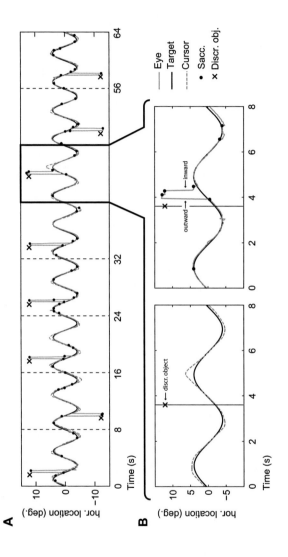

the pursuit target moved away (240 ms, $F(1,11) = 13.5$, $p < 0.01$, see also Fig. 14). The analysis of the discrimination object onset times showed shorter RTs for late (225 ms) and longer RTs for early onsets (237 ms, $F(1,11) = 5.9$, $p < 0.05$). Interactions between all factors were not significant. This suggests that the influence of the motion direction was independent from the effect of onset time.

The eccentricity of the target at saccade onset may affect the saccade RT (Kalesnykas and Hallett, 1994). To consider this possibility, we compared the amplitudes of the saccades. This showed that saccades were slightly larger when the pursuit target moved to the discrimination object (13.1°) and smaller when it moved away from the object (11.8°, $F(1,11) = 128.3$, $p < 0.01$).

A reduction of SRTs may be accompanied by a reduction in the accuracy of the saccades (Fischer et al., 1993). Our results show a difference in saccade gain between both motion conditions. Saccades that were initiated while the discrimination object appeared in motion direction (to condition) exhibited a lower gain (0.99) than saccades to discrimination objects at the opposite location (away condition, 1.02, $F(1,11) = 6.2$, $p < 0.05$). A gain greater than one indicates overshoot, whereas a gain smaller than one indicates undershoot. To examine how this difference in saccade gain translates to fixation accuracy, we compared the magnitude of the error between the saccade endpoint and the discrimination target's location. The results show no

---

Figure 13 *(preceding page)*: **A** Time course of stimulus presentation and response for a representative series of trials (8 trial epochs, 8 seconds each). **B** Close-up of one trial epoch. Left: Sinusoidal motion of the pursuit target and time of onset (here 3.6 s into the trial, in the direction of motion) of the discrimination object. The cursor motion shows a slight overshoot when the pursuit target's motion reaches its maximum and a pronounced overshoot after the discrimination object was presented. Right: The Gaze movements during the same trial show periods of smooth pursuit, small catch-up saccades, and large saccades to the discrimination object (outward saccade, ca. at 4 s) and back to the pursuit target shortly afterward (inward saccade).

significant difference between both conditions (average absolute error 0.45 °).

Depending on the SRT after onset of the discrimination object, outward saccades were either initiated before or after the eye crossed the display midline. An analysis of the starting position of outward saccades showed that saccades were initiated before the eye crossed the midline in the majority of cases (83%). Average SRTs between both motion conditions were compared for SRTs shorter than 400 ms (before midline crossing). The main SRT results also hold for this subset: saccades that were initiated while the pursuit target moved to the discrimination object exhibited shorter RTs (210 ms) compared to saccades that started when the pursuit target moved away (238 ms, $t(11) = 3.5$, $p < 0.01$).

Inward Saccades

Unlike saccades to the discrimination object, saccades back to the pursuit target from the discrimination object (inward) were not triggered by an experimental signal (i.e., go signal or stimulus onset) but on participants own initiative following the discrimination task. Since we were interested in influences of the movement direction of the pursuit target on saccade performance we first verified if our ex-ante classification of movement direction (to/away) was valid also for inward saccades. This was necessary to test if participants waited to saccade back after the pursuit target reached its maximal amplitude and changed direction. The analysis of inward saccade onsets showed that this was not the case. On average, and for the majority of trials (99.7%), saccades back to the pursuit target were initiated before the pursuit signal changed its direction (825 ms after discrimination object onset on average). This means that the classification, which was based on the experimental manipulation into saccades that were initiated while the pursuit target moved to or away from the discrimination object, was correct for the majority of trials.

Saccades back to the pursuit target from the discrimination object (inward) took much longer than outward saccades (overall mean SRT: 525 ms). Note that this time was measured from fixation onset on the discrimination object and therefore also comprised the dis-

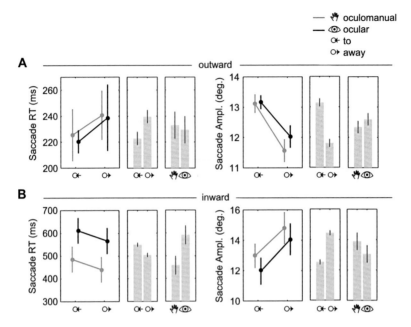

Figure 14: Saccade RTs and amplitudes of saccades after onset of the discrimination object (outward) and back to the pursuit target (inward) depending on task (oculomanual or ocular pursuit) and motion of the pursuit target relative to the discrimination object (to or away). Plots show task-motion means with standard deviations (centered on participant-means). Bar charts show means of individual factors with 95% confidence intervals. **A** Outward saccade RTs were shorter when the discrimination object was presented in the direction of the pursuit target's motion. Amplitudes were larger when the pursuit target moved to the discrimination object. **B** Inward saccades were initiated earlier during the oculomanual condition and were shorter when the pursuit target moved away. Amplitudes were larger in the away condition.

crimination time. Reaction times of inward and outward saccades can therefore not be compared directly. The ANOVA results show a statistically significant difference in SRTs for the pursuit task type ($F(1,11) = 13.9$, $p < 0.01$) and the motion direction of the pursuit target ($F(1,11) = 24.0$, $p < 0.01$). Saccades back to the pursuit target were initiated earlier during oculomanual pursuit (460 ms) compared to ocular pursuit (589 ms). SRTs were shorter when the pursuit target moved away from the discrimination object (503 ms) and longer when the pursuit target moved to the discrimination object (541 ms).

Amplitudes of inward saccades showed a significant main effect of motion direction ($F(1,11) = 169.0$, $p < 0.01$). Saccades were shorter when the pursuit target moved to the discrimination object (12.5 °) and longer when it moved away (14.4 °). The analysis of inward saccade gain showed no significant differences (average gain 0.992).

Discrimination task performance did not differ significantly between the pursuit conditions and also not between motion directions (on average 88% correct).

Pursuit Eye-Movements

The quality of smooth pursuit eye movements was measured by counting the number of saccades per second and in the form of the RMS error and phase shift between the eye and pursuit target. Periods during which the discrimination object was present were removed prior to calculating these measures. The results show that the mean number of saccades per second was higher during ocular pursuit (1.8 saccades/s) and lower during oculomanual pursuit (1.4 saccades/s, $t(11) = 4.5$, $p < 0.01$, Fig. 15 B). A larger RMS error between the eye and pursuit target was measured for ocular pursuit (1.20 °) in comparison to oculomanual pursuit (0.84 °, $t(11) = 3.1$, $p < 0.05$). The frequency spectrum of the eye movements exhibited maximal power at 0.25 Hz, which was the frequency of the target signal. The average phase shift between eye and pursuit target was lower in the ocular (-0.5 °) than in the oculomanual condition (3.6 °, $t(11) = 2.5$, $p < 0.05$). This corresponds to a lead of 5.5 ms in the ocular and a lag of 40 ms in the oculomanual condition. The phase shift of the eye was negative in 46% of blocks in the ocular condition but only in 5% of blocks in the oculomanual condition. Thus, in almost half of the measurements, the eye did not follow but precede the pursuit target in the ocular condition (see also Fig. 15 B, phase shift). The regression of block phase shift and average number of saccades per second showed no significant correlation for the oculomanual condition (-0.004 saccades/s) but a negative slope for the ocular condition (-0.03 saccades/s, $t(58) = 2.7$, $p < 0.01$, $r = 0.34$). This suggests that anticipation of the target's trajectory is accompanied by a moderate increase in saccade frequency in this condition.

3.5 DISCUSSION

In the current study we examined how smooth pursuit eye movements influenced task-switching saccades. Participants alternated their gaze between a continuous pursuit and a discrete object discrimination task. The main results of our study show asymmetries in saccade reaction times (SRTs) from and to smooth pursuit depending on the smooth pursuit movement direction.

Outward SRT Asymmetry

We examined whether ongoing pursuit influenced initiation of *outward* saccades (saccades from the pursuit to the discrimination task). The SRTs of outward saccades were shorter when the saccade target appeared in the direction of the pursuit target's movement.

In explaining this result, we first address two basic factors that are known to influence SRTs, namely the eccentricity of the saccade target and the orbital position of the eye at saccade onset. SRTs have been shown to be a function of target eccentricity (Kalesnykas and Hallett, 1994, but see also Hodgson, 2002; Dafoe et al., 2007). For example, in the study by Kalesnykas and Hallett (1994) longer SRTs were measured for extremely small ($< 2\,°$) and very large eccentricities ($> 15\,°$). Our analysis of saccade amplitudes indicates that eccentricities at the time of saccade onset differed in the two motion conditions. Saccades in the direction of motion (to condition) were larger than saccades in the opposite direction. However, considering the pattern of results of Kalesnykas and Hallett (1994), longer rather than shorter SRTs

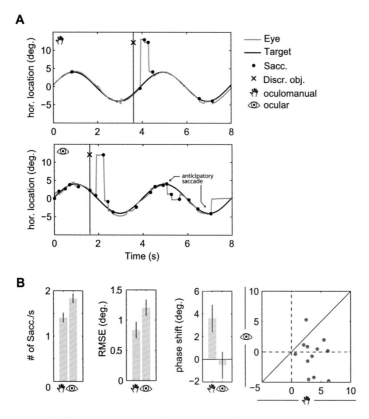

Figure 15: Differences in smooth pursuit behavior between ocular and oculomanual pursuit. **A** Traces of eye movements during oculomanual (top) and ocular pursuit (bottom) exemplify interruptions of smooth pursuit by small saccades that sometimes move gaze away from the target. **B** Left: Bar charts show average pursuit metrics with 95% confidence intervals. The number of saccades increased and RMS error increased during ocular pursuit. On average, the eye lagged behind the target during oculomanual pursuit but preceded it slightly during ocular pursuit. Right: The scatterplot of participant phase shift means shows that nearly half of the means were negative in the ocular condition but none was negative in the oculomanual condition.

would be expected for larger eccentricities. Another factor that directly affects SRTs is the orbital position of the eye. It has been shown that centripetal saccades, saccades from an eccentric starting positions that move the eyes back to the primary position, exhibit shorter SRTs than saccades in the opposite condition (Fuller, 1996; Paré and Munoz, 2001). In the current experiment, orbital positions were approximately equal (namely close to the primary position) at saccade onset. This suggests that the observed SRT asymmetries are neither linked to eye or target position but to the target's motion direction.

This observation is in line with previous research that explains asymmetric SRTs on the basis of an attentional bias in the direction of pursuit (Tanaka et al., 1998; Kanai et al., 2003; Khan et al., 2010). For example, Khan et al. (2010) showed shorter SRTs to targets ahead of the pursuit stimulus and longer SRTs to targets behind the pursuit target. This phenomenon has also been reported for manual response times in a detection task (van Donkelaar, 1999; van Donkelaar and Drew, 2002). This suggests that SRT asymmetries are not due to biomechanical compatibility between saccade and pursuit direction but rather an example for an attention shift in the direction of pursuit. Khan et al. (2010) suggest that orienting of covert attention in anticipation of the pursuit target's motion is important such that potentially required actions can be planned ahead, in compensation for neural processing delays in perception and action.

The results of our study exclude a basic explanation for this attentional bias. In pursuit tasks that do not require a visual analysis of the target of some sort, the pursuit behavior is not completely smooth but shows discontinuities in the form of anticipatory saccades (Van Gelder et al., 1990, see also Koken and Erkelens, 1992; Xia and Barnes, 1999). According to Van Gelder and colleagues (Van Gelder et al., 1990, 1995b,a; Kathmann et al., 1999), pursuit is typically performed automatically to support the visual analysis of the target. Without such a visual function, attention is unnaturally focused on pursuit itself, which may explain any anticipatory behavior. Our hypothesis was that this tendency to anticipate the pursuit target's motion could have also caused the attentional bias and reduction of SRTs in the motion direction. However, the current results speak against this assumption. SRTs were asymmetric in both pursuit conditions (ocular, oculomanual), despite clear differences in smooth pursuit behavior in regard to anticipation (see also Mather and Putchat, 1983; Gauthier et al., 1988; Vercher and Gauthier, 1992; Koken and Erkelens, 1992). This suggests that pursuit-related attentional biasing is not merely the result of confined experimental settings and extends its relevance to more realistic conditions.

Inward SRT Asymmetry

We tested whether SRT asymmetries also existed for *inward* saccades (saccades from the discrimination object to the pursuit target). SRTs were shorter when the saccade target moved away from the current fixation location (foveofugal) and longer when it moved toward the fixation location (foveopetal).

Like outward saccades, SRTs were shorter when the saccade moved the eye in the same direction as the pursuit target. An advantage for saccades that are compatible with the pursuit motion direction was explained for saccades *from* pursuit by a broad attentional bias in the direction of pursuit, which facilitates detection and processing of targets that appear in this direction (Blohm et al., 2005; Khan et al., 2010). An alternative explanation suggests that it is not a sustained bias in attention but facilitation of attention capture, which leads to reduced SRTs to sudden target onsets in the direction of pursuit (Lovejoy et al., 2009).

Neither theory sufficiently explains the current results. First, when fixating on the discrimination object prior to the inward saccade, pursuit targets were situated in the same visual hemifield at similar visual field locations in both motion conditions. A broad tuning of attention would therefore be expected to affect saccades in both conditions. Second, inward saccades were not triggered by a sudden target onset. Instead, saccades followed the discrimination task and moved the eye to the pursuit target, which was continuously present throughout the experiment. Hence, facilitation of attention capture is also unlikely to explain the obtained result.

In the remainder of this discussion we will consider several alternative explanations, namely the influence of the discrimination task, amplitude differences, motion processing asymmetries, compatibility with early pursuit responses, and inhibition of saccades at the onset of smooth pursuit.

The amount of time spent on performing the discrimination task may explain differences in SRTs. For example, longer discrimination times may be the result of inaccurate foveation after the outward saccade. However, our analysis provides no evidence for this assumption. Discrimination performance and saccade accuracies were similar in both conditions. In addition, the current finding is corroborated by data from a different experiment, in which neither discrimination nor a saccade was required before the saccade to pursuit (Bieg et al., 2013b, in preparation).

Factors that influence SRTs more directly are the eccentricity of the saccade target and the orbital position of the eye at saccade onset. However, the eccentricity differences in our experiment would predict the opposite effects on SRTs (see previous section). This suggests that the observed SRTs are primarily influenced by the motion direction of the pursuit target.

Asymmetries in the processing of motion have been observed in several experiments. But the conditions that would lead to an advantage in one or the other direction (foveofugal/foveopetal) are not clear (Naito et al., 2010). For example, in an experiment by Ball and Sekuler (1980), RTs to motion onsets of foveofugal motion were shorter. Other experiments showed an advantage for foveopetal motion (Mateeff and Hohnsbein, 1988; Mateeff et al., 1991b,a; Raymond, 1994; Jancke et al., 2004). One reason for these conflicting findings could be differences in the presented type of motion. Mateeff et al. (1991b) compared flow-field motion (i.e., random-dot kinematograms) stimuli and single-target motion stimuli. The latter stimulus is similar to the one that was used in the present experiment. Mateeff et al. (1991b) show that stimuli of this sort lead to processing advantages of foveopetal motion (in terms of onset detection) rather than foveofugal motion, as in our experiment (in terms of SRTs). These findings speak against an explanation on the basis of motion processing asymmetries.

Potentially related to asymmetries in motion processing are asymmetries in smooth pursuit behavior. These can be observed during the early (ca. 100 ms), open-loop pursuit response (Tychsen and Lisberger, 1986; Carl and Gellman, 1987). This response can occur at the onset of pursuit and moves the eyes in the direction of the pursuit target's motion. Investigations of this response showed larger early accelerations during foveopetal motion (Tychsen and Lisberger, 1986). This initial acceleraion could potentially affect saccade onsets by modulating the omnipause neuron activity in the brain stem. Inhibition of these neurons is required to trigger a saccade (Scudder et al., 2002) and they also likely regulate smooth pursuit onset and gain (Missal and Keller, 2002; Kornylo et al., 2003; Krauzlis, 2005). Inhibition of omnipause activity due to early pursuit responses could therefore facilitate saccade triggering. With regard to the findings by Tychsen and Lisberger (1986), stronger inhibition of omnipause neurons would be expected when the target moves foveopetally, which would explain shorter SRTs in this direction. Again, this is incompatible with the results that we observed, namely shorter SRTs to foveofugal motion.

Apart from this hypothetical facilitatory connection, pursuit-related activity is known to *inhibit* saccades in certain conditions. Increased SRTs or even complete suppression of a saccade can be observed in foveopetal step-ramp tasks. There, the target is stepped in the opposite movement direction such that it moves across its original position after a specified time. This time is the *zero-crossing* or *eye crossing* time (Gellman and Carl, 1991; de Brouwer et al., 2002a). In the case of zero-crossing times of 200 ms, the initial saccade to the target position is delayed or suppressed completely and smooth pursuit of the target commences directly (Rashbass, 1961; Gellman and Carl, 1991). It is currently unknown how this cancellation process affects saccades for zero-crossing times larger than 200 ms. For example, the study by Moschner et al. (1999) measured SRTs in step-ramps with 200 ms zero-crossing times. Their results show longer SRTs in foveopetal steps (ca. 400 ms) and shorter SRTs in foveofugal steps (ca. 200 ms). However, this difference in SRTs can be primarily attributed to cancellation of the initial saccade and re-planning of a new saccade in the direction of motion after zero-crossing. In contrast, SRT differences in inward saccades in our experiment can not be attributed to cancellation and re-planning since (1) zero-crossing never actually occurred and (2) hypothetical zero-crossing times were much longer: An estimate based on the average amplitude prior to saccade onset (12.5°) divided by the pursuit target speed (max. 6.7°/s, average before onset 5.2°/s) results in zero-crossing times between 1.8 and 2.4 s.

It cannot be excluded that the same mechanisms that lead to cancellation of saccades in short zero-crossing times also influence saccade generation in longer zero-crossing times. Saccade triggering as well as cancellation are thought to depend on neuronal accumulation processes (Carpenter and Williams, 1995; Hanes and Schall, 1996). Importantly, there is also evidence for inhibitory links between those processes (Boucher et al., 2007). Assuming that cancellation of saccades to foveopetal motion is indeed organized by such a process network, foveopetal motion would be expected to have a stronger impact on the cancellation process gain than foveofugal motion. The inhibitory connections between the two processes can then explain increased SRTs to foveopetal motion. In this respect it is important to point out that an asymmetry in SRTs may also be behaviorally useful. In foveofugally moving targets, computation of the exact time of zero-crossing from a motion analysis of the pursuit target becomes obsolete. Considering that a more precise motion estimate also requires more time (Bruyn and Orban, 1988; Bennett et al., 2007), it would be efficient to allocate less time for the analysis of foveofugal rather than foveopetal motion. In particular because foveofugal motion moves the target out of the visual field, which poses the danger of losing track of it entirely when the saccade is triggered too late. In this respect, the ensuing reduction in saccade RTs may additionally be related to time pressure (Reddi and Carpenter, 2000; Montagnini and Chelazzi, 2005; Bieg et al., 2012).

3.6 CONCLUSION

We examined how smooth pursuit eye movements influenced initiation of saccades in the context of task-switching. Here, gaze had to be switched from and to a pursuit task. First, our results confirm earlier findings which show that the relative movement direction of the pursuit stimulus affects saccade reaction times (SRTs) *from* pursuit. Our results also provide evidence against a potential explanation for this behavior, namely the tendency to anticipate the pursuit target's trajectory, which is particularly pronounced in basic, laboratory pursuit tasks (Van Gelder et al., 1995a).

Second, our results show that saccades *to* pursuit are similarly affected by the relative movement direction of the pursuit target. We speculate that the difference in SRTs may be caused by the processes that organize cancellation of saccades at the onset of pursuit movements (Rashbass, 1961; Gellman and Carl, 1991). Additional studies are required to establish the exact conditions, for example, changes in the zero-crossing time (de Brouwer et al., 2002b), that lead to these SRT differences. This would allow a more precise specification to which extent saccades are influenced when switching to pursuit behavior.

ACKNOWLEDGEMENTS

We thank Dr. Frank Nieuwenhuizen for his comments and suggestions. This research was supported by the Max Planck Society, by the myCopter project, funded by the European Commission under the 7th Framework Program, and by the WCU (World Class University) program funded by the Ministry of Education, Science and Technology through the National Research Foundation of Korea (R31-10008).

COPYRIGHT

This article is distributed under the terms of the creative commons attribution license which permits any use, distribution, and reproduction in any medium, provided the original author(s) and the source are credited.

4

COORDINATION OF SACCADES AND SMOOTH PURSUIT

This chapter has been reproduced from an article that was submitted for publication: Bieg, H.-J., Chuang, L. L., Bülthoff, H. H., & Bresciani, J.-P. (2013). Asymmetries in saccade reaction times to pursuit.

4.1 ABSTRACT

Do saccade reaction times (SRTs) of saccades to moving objects depend on the movement direction of the object? Participants performed a step-ramp task in which the target object stepped from a central to an eccentric position and moved at constant velocity either to the fixation position (foveopetal) or further to the periphery (foveofugal). The step size and target speed were varied. Observers' responses were categorized and trials were selected that exhibited a saccade prior to a smooth pursuit eye movement. These saccades moved the eyes either in the direction of pursuit in foveofugal trials or in the opposite direction in foveopetal trials. The results show longer SRTs in the foveopetal condition. In this condition, SRTs depend on the target speed and step size. Consistent with this, the occurrence of an initial saccade and strength of pre-saccadic accelerations also depend on the target speed and step size in foveopetal trials. A common explanation for these results may be found in the mechanisms that select between oculomotor response alternatives (i.e., a saccadic or smooth response).

4.2 INTRODUCTION

Saccades are rapid movements of the eye that move the retinal image of an object to the fovea, the area of highest acuity (Carpenter, 1988; Gilchrist, 2011). Two processes are thought to control saccades. First, those that determine the endpoint location of the saccade (where) and second, processes that determine the time of saccade onset (when; Findlay and Walker, 1999; Becker and Jürgens, 1979).

In saccades to moving objects, the coordination of both decisions is very important since the object's position undergoes a continuous change during the time period leading up to the saccade onset. Several studies have investigated saccades to moving objects, in particular, addressing the *where* aspect of saccadic control. This has shown that saccades are programmed to compensate for target movements during saccade preparation by considering the target's velocity (Keller and Johnsen, 1990; Gellman and Carl, 1991; Kim et al., 1997; Eggert et al., 2005b,a; Guan et al., 2005; de Brouwer et al., 2002a; Etchells and Benton, 2010; but see Heywood and Churcher, 1981; Smeets and Bekkering, 2000). For example, in a recent study by Etchells and Benton (2010), participants performed saccades to targets that moved horizontally at varying speeds. Their analysis showed that saccade endpoint error was best explained by a model that incorporates both the target's position and the velocity of the target 100–300 ms before saccade onset.

The *when* aspect of saccades to moving stimuli has primarily been investigated in the context of smooth pursuit eye movements (reviewed by Ilg, 1997; Krauzlis, 2005; Thier and Ilg, 2005; Barnes, 2008). Here, the conditions have been examined that determine the presence or absence of an initial saccade at the beginning of smooth pursuit. When following a moving object, its retinal image is first foveated by an initial saccade and then stabilized by a smooth movement of the eye at matching velocity (Lisberger, 1998). However, a saccade at the beginning of smooth pursuit is not always present. Its occurrence depends on the *zero-crossing* (Gellman and Carl, 1991) or *eye crossing* time (de Brouwer et al., 2002a). For example, smooth pursuit commences directly when the target object crosses the observer's current fixation location within approximately 200 ms (Rashbass, 1961). It is important to note that cancellation of a saccade has been estimated to take approximately 100 ms. For zero-crossing times > 100 ms these findings therefore suggest that the decision to suppress a saccade to a moving target is based on an extrapolation of the target's position (Gellman and Carl, 1991). The question arises whether this extrapolation mechanism affect the final trigger decision or whether it exerts a more general influence on the *when* aspects of saccade preparation.

SRTs to moving targets have been investigated by Gellman and Carl (1991) and Moschner et al. (1999). Both studies showed that saccades following steps in the opposite pursuit movement direction (foveopetal) exhibit longer SRTs than steps in the direction of pursuit (foveofugal). Unfortunately, in both studies, target speed and step amplitudes were selected close to zero-crossing times of 200 ms. The consequence of this is a suppression of the initial saccade at the onset of pursuit (Rashbass, 1961). In the cited studies, the prolonged SRTs therefore arose from saccades that occurred after zero-crossing, namely catch-up saccades in the direction of motion. Therefore, the reported difference in SRTs between foveopetally and foveofugally moving targets may not necessarily reflect a differential effect on saccade preparation but could be regarded as a direct result of saccade cancellation and re-planning instead.

Recent results from our laboratory suggest that processing of the target's motion in saccades to moving stimuli may generally affect the time point of saccade onsets also independently from such cancellation delays (Bieg et al., 2013a). In an experiment that required observers to move their gaze from and to a smoothly moving stimulus, longer saccade reaction times (SRTs) were measured when the saccade moved the eyes in the opposite direction of pursuit. An exact computation of the zero-crossing time in this previous study was not possible because the pursuit target moved on a sinusoidal path, which only approached but never reached the observers' fixation location before the saccade. Furthermore, saccades also differed systematically in amplitude due to the experimental design.

In the current study we examine the conditions that lead to asymmetric SRTs at the onset of pursuit without these limitations. Specifically, we examine whether and how SRTs are affected by target eccentricity, speed, and zero-crossing time prior to saccade onset. For this, SRTs were measured to moving stimuli in a horizontal step-ramp paradigm. We manipulated the step size and target speed to induced different zero-crossing times. We chose zero-crossing times between 100 ms and 1.2 s such that it was possible to measure the SRTs of initial saccades before zero-crossing rather than catch-up saccades after zero-crossing. A categorization of responses was performed to select saccades in the direction of pursuit or against the pursuit direction. The SRTs of these saccades were subsequently compared.

Participants

26 participants took part in the experiment, 10 in the 20°/s condition (5 male, 5 female, age: 23–29 years) and 16 in the 10°/s condition (7 male, 9 female, age: 19–31 years). All participants had normal or corrected to normal vision. A vision test was conducted to verify this prior to the experiment (FrACT test Bach, 2007, logMAR < 0 for all participants). In accordance with the World Medical Association's Declaration of Helsinki, written informed consent was obtained from all subjects prior to experimentation. Participants were paid 8 EUR per hour for taking part in the experiment.

Materials

Participants sat in an adjustable chair in front of a TFT monitor (Samsung 2233RZ, 120 Hz refresh rate, resolution 1680 × 1050, see also Wang and Nikolić, 2011). A chin-rest provided support for the head at a viewing distance of 57 cm. An optical infrared head-mounted eye-tracking system was used to measure gaze at a sampling rate of 500 Hz (SR Research Eyelink II).

Task

Participants followed the horizontal motion of a pursuit target that was shown on the computer screen as closely as possible (see Fig. 16). The pursuit target consisted of a disk that was rendered with a smooth, circular gradient from white to gray (RGB 255, 255, 255; to RGB 100, 100, 100) and subtended 0.8°. Throughout the experiment a uniform gray background (RGB 100, 100, 100) was presented. At the beginning of a trial, the disk appeared at the center of the display. The disk then stepped either to the left or right after a random delay between 1–2 s. The amplitude and direction of the step was selected randomly. Six different amplitudes from 2° to 12° (in steps of 2°) were presented. After the step, the disk moved with a constant velocity of 20°/s or 10°/s either toward the observer's fixation location (foveopetal) or away from it (foveofugal). One out of 25 trials was randomly designated a catch trial in which no target step occurred.

Design and procedure

Target-speed was varied between subjects. The experiment was run with target speeds of $20\,°/s$ for 10 participants and $10\,°/s$ for 16 participants. Within-subjects factors were the movement direction after the target step (foveopetal, foveofugal) and the target step amplitude (six amplitudes, see above).

In foveopetal trials, this resulted in different zero-crossing times. This is the time that the target requires to reach its original (zero) position after the step. The presented step amplitudes resulted in zero-crossing times from 100 to 600 ms (in steps of 100) for target speeds of $20\,°/s$ and 200 to 1200 ms (in steps of 200) for target speeds of $10\,°/s$.

During a session, tasks were presented in several runs. Each run took ca. 15 minutes including set-up and calibration of the eye-tracker. During a run, participants performed 5 blocks of the experimental task with 25 trials each. Regular breaks were provided after each run, during which the eye-tracker was removed. The entire experimental session lasted ca. 120 minutes.

Data analysis

Saccade detection was carried out by the Eyelink II system using a velocity ($22\,°/s$) and acceleration threshold ($3800\,°/s^2$).

The primary measures used to characterize saccadic eye movements were saccade reaction time (SRT) and saccade amplitude. SRT was defined as the time between the onset of the target step and initiation of the saccadic movement.

Data from the following trials were removed prior to saccade analysis: Trials with blinks during the critical time period shortly before or after the target step, missed trials (no saccade or RT greater than 800 ms), anticipatory saccades (RT smaller than 50 ms), and inaccurate saccades with errors larger $2\,°$ visual angle. Based on this method 7.5% data points were removed in the $20\,°/s$ condition and 5% in the

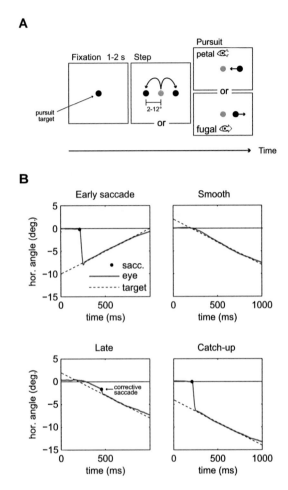

Figure 16: **A** Schematic of the experimental task. Participants fixated a disc at the center of the screen. The disc randomly stepped to the left or right of the display. After the step, the disc moved at a constant velocity either away from the center (foveofugal) or towards the center (foveopetal). **B** Observed responses. In foveopetal trials, observers either performed an early saccade against the motion direction before it crossed the fixation point, directly initiated a smooth pursuit movement, or caught-up with the target after it crossed the fixation point. In foveofugal trials, participants always performed a catch-up saccade.

10°/s condition. The median number of data points remaining per participant, velocity, step amplitude, and motion direction condition was 40 (min. 21).

If not indicated otherwise, data plots show Cousineau-Morey confidence intervals (see Baguley, 2012; Morey, 2008).

4.4 RESULTS

Response Types

A typical response to a target step and foveofugal motion is shown in Fig. 16 A (bottom right). Here, a catch-up saccade is initiated after a latency of approximately 200 ms. In trials with foveopetal motion (toward the fovea) three main types of responses can be observed: (1) an early saccade against the movement direction before the target crosses the fixation location, (2) direct initiation of smooth pursuit without an initial saccade, or (3) a late saccade after the target crossed the fixation location in the direction of movement (see Fig. 16 B). We collapsed smooth and late trials because a distinction would solely be possible on the basis of an arbitrary temporal threshold and is not of interest to the question at hand.

The proportion of these responses is expected to depend on the zero-crossing time (de Brouwer et al., 2002b) and thus on the step amplitude and speed in the current experiment. Smooth and late saccade trials are expected to occur when the target reaches the zero position early, that is, for small step amplitudes and high speeds. Early saccades are expected to occur when the target reaches the zero position late, i.e., for large amplitudes and low speeds. To verify this, the proportions of trial types were computed across all subjects and step amplitudes separately for each velocity condition.

Fig. 17 A shows histograms of observed responses. The results show mainly smooth/late responses for short step amplitudes and mainly early responses for large step amplitudes (see Table 1). The proportions show that early responses constitute the major response type ($>$ 50%) for steps equal to or larger than 8° in the 20°/s condition and 4° in the 10°/s condition.

Table 1: Average proportion of saccade responses (percent) by step amplitude. See also histogram Fig. 17 top

		2°	4°	6°	8°	10°	12°
20°/s	early	0.3	0.4	25.7	77.9	94.0	98.8
	late	99.7	99.6	75.3	22.1	6.0	1.2
10°/s	early	1.6	82.7	99.7	99.5	100.0	99.5
	late	98.4	17.3	0.3	0.5	0.0	0.5

Saccade Reaction Times

We compared saccade reaction times (SRTs) of early responses with the SRTs of catch-up responses. Early saccades move the eye against the motion direction of the pursuit target, catch-up saccades move the eye in the direction of the pursuit target (see Fig. 16 B). SRTs were only compared between conditions in which the proportion of early saccades constituted the majority in foveopetal trials. Furthermore, SRTs of both velocity conditions were analyzed separately because of the different distributions of responses between both conditions (see previous section). The results are presented in turn.

For trials in the 20°/s condition, a 2 × 3 repeated-measures ANOVA was conducted. The tested factors include the motion direction of the pursuit target (foveopetal, foveofugal) and three step amplitudes (8°, 10°, 12°). These were the amplitudes where the predominant saccade response was of the "early" type in foveopetal trials ($> 50\%$). Fig. 17 B (left) shows the average SRTs and tested conditions. The analysis shows significantly longer SRTs in the foveopetal condition (236 ms) in comparison to the foveofugal condition (190 ms, $F(1,9) = 34.5$, $p < 0.01$).

For trials in the 10°/s condition, the range of step amplitudes that were considered in the analysis was extended from 4° to 12° (see Fig. 17 B, right). Again, these were the amplitudes where the predominant saccade response was of the "early" type in foveopetal trials. A 2 × 5 repeated-measures ANOVA shows significant longer SRTs in the foveopetal condition (209 ms) in comparison to the foveofugal condition (186 ms, $F(1,15) = 51.9$, $p < 0.01$). For foveopetal trials, a linear regression of step amplitude and SRTs shows a negative correlation ($-3.8\,\text{ms}/\,^\circ$, $t(78) = 3.0$, $p < 0.01$, $r = -0.32$). In other words: SRTs decreased with increasing step amplitude for these trials.

Amplitudes and Eccentricity-Matched SRTs

Saccades to targets at different eccentricities could also systematically differ in SRTs (Kalesnykas and Hallett, 1994). Differences in eccentricities at the onset of the saccade and thus also differences in the amplitudes of the saccades were expected in the current experiment due to the motion of the target during preparation of the saccadic response (i.e., the SRT). Saccades in the $20\,^\circ/\text{s}$ condition were smaller in foveopetal step ramps ($5.0\,^\circ$) than in foveofugal step ramps ($13.6\,^\circ$, $F(1,9) = 427$, $p < 0.01$). Similar results are obtained in the $10\,^\circ/\text{s}$ condition (foveopetal amplitudes $5.7\,^\circ$, foveofugal amplitudes $10.0\,^\circ$, $F(1,15) = 847$, $p < 0.01$).

Fig. 17 C shows average saccade amplitudes per step condition. As can be seen from this graph, a shift of the diagonal iso-amplitude line according to the expected amplitude difference for a 200 ms saccade reaction time approximates the measured amplitudes well for both pursuit target speeds. This corroborates our assumption that the movement of the target during saccade preparation caused the obtained amplitude differences.

To test whether the obtained differences in SRTs between saccades in foveofugal and foveopetal trials were independent of the target's eccentricity at saccade onset, a comparison of eccentricity-matched conditions was conducted. For $20\,^\circ/\text{s}$ trials, the selected conditions were $2\,^\circ$ and $4\,^\circ$ steps for foveofugal trials and $10\,^\circ$ and $12\,^\circ$ steps for foveopetal trials. Assuming a constant SRT delay of 200 ms the targets in these two conditions were approximately at $6\,^\circ$ and $8\,^\circ$ eccentricity at saccade onset, respectively (see Fig. 18 A). A 2×2 repeated-measures ANOVA was conducted to compare average SRTs. The results show significantly longer SRTs in the foveopetal condition (232 ms) than in the foveofugal condition (198 ms, $F(1,9) = 13.4$, $p < 0.01$). To assess whether the difference in SRTs is accompanied by a difference in saccade accuracy we compared the absolute saccade endpoint error

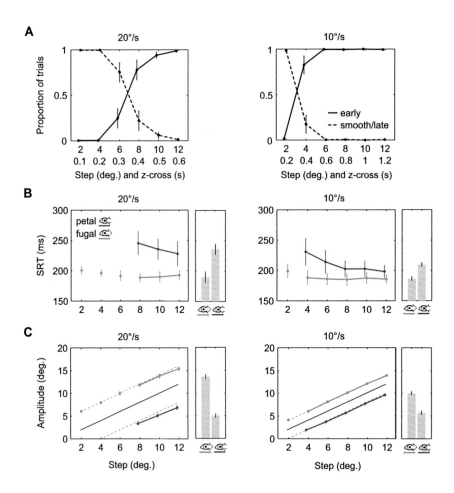

between both motion conditions. This showed no significant difference (average absolute error 0.65 °).

For 10 °/s trials, the selected conditions were 2, 4, 6, and 8 ° steps for foveofugal trials and 6, 8, 10, and 12 ° steps for foveopetal trials. The targets in these conditions were approximately at 4, 6, 8, and 10 ° eccentricity at saccade onset, respectively (see Fig. 18 A). A 2 × 4 repeated-measures ANOVA was conducted to compare average SRTs. Again, the results show significantly longer SRTs in the foveopetal condition (204 ms) than in the foveofugal condition (189 ms, $F(1, 15) = 43.2, p < 0.01$). Differences in absolute saccade endpoint error were not significant (average absolute error 0.34 °).

Target Speed

To analyze the effects of target speed on SRTs we first compared differences between conditions with equal target eccentricities at the time of the approximate saccade onset 200 ms after the step. For 20 °/s foveofugal trials, target eccentricities are 6, 8, 10, and 12 ° for steps of 2, 4, 6, and 8 °, respectively. Matching 10 °/s trials have steps of 4, 6, 8, and 10 °. The comparison shows no statistical difference between

---

Figure 17 *(preceding page)*: **A** Proportion of foveopetal trials with early saccades (against the motion direction) and smooth/late trials. The latter comprised of trials with late saccades (in the motion direction) or trials without an initial saccade and direct initiation of smooth pursuit. Error bars show standard deviations. **B** Average saccade reaction time (SRT) per step condition. Averages only include "early" saccades in the foveopetal case. Data points for conditions where the proportion of these responses was < 50% are omitted. Connected data points show conditions that were subjected to an analysis of variance. The bar charts to the right of each graph show the average SRTs for each motion condition. **C** Average saccade amplitude per step condition. The black line shows amplitudes matching the step size. The broken lines depict the target's position after 200 ms considering the respective velocity. Error bars show standard deviations (linecharts) and 95% CIs (barcharts).

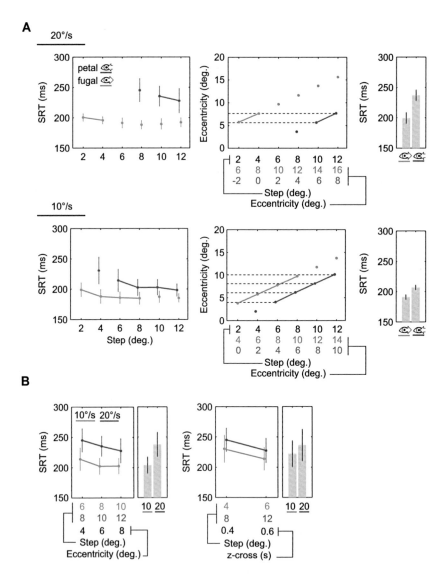

SRTs between both speed conditions for foveofugal trials (average SRT 190 ms). This suggests that target speed has no influence on the saccade onset timing in saccades to foveofugal targets.

For $20°/s$ foveopetal trials, target eccentricities are 4, 6, and $8°$ for steps of 8, 10, and $12°$, respectively. Matching $10°/s$ trials have steps of $6°$, $8°$, and $10°$ (see Fig. 18 B, left). The comparison shows significantly longer SRTs in $20°/s$ trials (248 ms) than in $10°/s$ trials (206 ms, $F(1,24) = 5.4$, $p < 0.05$).

However, the current data does not show a statistical distinction between both speed conditions for foveopetal trials when conditions are matched by zero-crossing time, which captures the combined influence of both speed and eccentricity (see Fig. 18 B, right). For $20°/s$ trials, conditions that result in zero-crossing times of 0.4 ms and 0.6 ms have target steps of $8°$ and $12°$. Matching $10°/s$ trials have steps of $4°$ and $6°$. The difference between both speed conditions is not statistically significant (average SRT 227 ms). This suggests that the zero-crossing time is the most useful predictor of SRTs in foveopetal trials.

Figure 18 *(preceding page)*: **A** Eccentricity-matched motion conditions. Top: $20°/s$ trials. Bottom: $10°/s$ trials. Left: SRTs of the conditions that were compared in the statistical analysis are connected. Middle: Eccentricities of targets shortly before saccade onset of the compared conditions. Eccentricities in the foveopetal and foveofugal case were approximately equal (see text for details). Theoretical eccentricities on the x-axis show the eccentricity of the target after the step and 200 ms of motion. Right: Difference between average SRTs in eccentricity-matched step amplitudes for the motion direction condition (foveofugal, foveopetal). **B** Conditions matched by eccentricity (left) and zero-crossing (right) for comparison of $20°/s$ and $10°/s$ trials. When matched by eccentricity, the two speed conditions show distinct SRTs. This is not the case when both conditions are matched by zero-crossing. Error bars show standard deviations (linecharts) and 95% CIs (barcharts).

Small, pre-saccadic accelerations (PSAs) in the direction of pursuit can sometimes be observed in saccades to moving targets (Fig. 19 A). To examine the occurrence and relationship of these eye movements with the measured SRTs we computed the average velocity and acceleration of the eye 50 ms prior to the saccadic eye movements.

First, average pre-saccadic velocities (PSVs) per condition were compared. As before, statistical analyses were performed for conditions in which the "early" response predominated in foveopetal trials. In the following report, positive velocities denote movements toward the pursuit target and negative velocities denote movements away from the pursuit target.

For trials in the $20°/$s condition, a 2×3 repeated-measures ANOVA was conducted. In both motion conditions, PSVs occur in the direction of pursuit. Specifically, in the foveopetal condition, the PSV moves the eye away from the target $(-2.8°/$s). In the foveofugal condition, the PSV moves the eye in the direction of the target $(0.9°/$s, $F(1, 15) = 34.6$, $p < 0.01$). For trials in the $10°/$s condition, a 2×5 repeated-measures ANOVA was conducted. Again, PSVs primarily occur in the direction of pursuit. The eye moves away from the target in the foveopetal condition $(-1.02°/$s) and in the direction of the target in the foveofugal condition $(0.93°/$s, $F(1, 15) = 39.3$, $p < 0.01$) before the initial saccade.

It should be noted that the actual displacement that is effected by these pre-saccadic accelerations is very small. On average, the displacement from the zero position was $0.15°$ for $20°/$s trials (max. $0.34°$) and $0.07°$ for $10°/$s trials (max. $0.19°$).

The same analysis of variance was repeated for pre-saccadic accelerations (PSAs, see Fig. 19 B). The results are in line with the PSV results: PSAs primarily move the eye in the direction of pursuit for both motion conditions. For target speeds of $20°/$s, the eye is accelerated away from the target in foveopetal trials $(-34°/$s^2) and toward the target in foveofugal trials $(15°/$s^2, $F(1, 9) = 51.7$, $p < 0.01$). Similar results are obtained for $10°/$s speeds. The eye is accelerated away from the target in foveopetal trials $(-17.5°/$s^2) and toward the target in foveofugal trials $(17.6°/$s^2, $F(1, 15) = 55.2$, $p < 0.01$). A regression of step amplitude and PSA shows a positive correlation in foveopetal

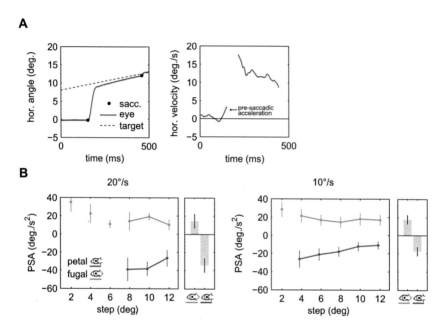

Figure 19: **A** Typical response during a foveofugal trial showing the slight pre-saccadic acceleration of the eye in the direction of pursuit prior to the saccade. Saccadic periods are omitted in the velocity plot. **B** Average pre-saccadic acceleration (PSA) per step condition. Averages only include "early" saccades in the foveopetal case. Data points for conditions where the proportion of these responses was < 50% are omitted. Connected data points show conditions that were subjected to an analysis of variance. Bar charts show the average SRTs for each motion condition. Error bars show standard deviations (linecharts) and 95% CIs (barcharts).

trials ($1.9°/s^2$ per deg., $t(78) = 3.8$, $p < 0.01$, $r = 0.4$). This means that accelerations are faster in trials with short target steps and slower in trials with large target steps.

A comparison of PSAs across the two speed conditions shows similar results to those of SRTs. Trials with equal target eccentricities at the time of the approximate saccade onset 200 ms after the step (see previous section for exact conditions) show significantly larger PSAs in the $20°/s$ speed condition ($-34°/s^2$) in comparison to the $10°/s$ condition ($-17°/s^2$, $F(1, 24) = 13.0$, $p < 0.01$). A statistical distinction is not possible when trials in both speed conditions are matched by zero-crossing time (average PSA $-27°/s^2$).

To examine whether the presence of these pre-saccadic effects affect SRTs, a subset of trials was analyzed with absolute PSVs smaller than $0.5°/s$. This analysis was only performed for $10°/s$ trials with target step amplitudes larger or equal than $6°$ to assure that data points were available for every participant and condition (33% of datapoints). The avarge PSV in this subset was $0.02°/s$ with an average PSA of $-1.5°/s^2$. A 2×4 repeated measures ANOVA shows significant differences in SRTs between both motion conditions also in this subset (Fig. 20). SRTs were longer in the foveopetal condition (196 ms) in comparison to the foveofugal condition (183 ms, $F(1, 15) = 16.36$, $p < 0.01$).

Furthermore, an analysis of this subset based on eccentricity-matched conditions as described before was performed. The selected conditions were 4, 6, and $8°$ steps for foveofugal trials and 8, 10, and $12°$ steps for foveopetal trials. Again, the compared conditions was limited to a range that assured that data points were available for all participants and conditions. A 2×3 repeated-measures ANOVA revealed significantly longer SRTs in the foveopetal condition (195 ms) than in the foveofugal condition (182 ms, $F(1, 15) = 14.1$, $p < 0.01$). This suggests that the presence of large pre-saccadic velocities or accelerations is not a precondition for the observed SRT asymmetries.

4.5 DISCUSSION

The current study examined saccade reaction time (SRT) differences to moving targets. The results show asymmetric SRTs: SRTs were

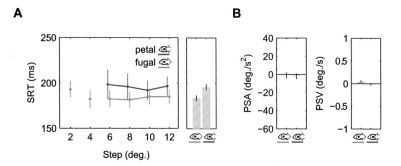

Figure 20: **A** SRTs of data with PSV $< 0.5\,^{\circ}/\mathrm{s}$. Connected data points show conditions that were subjected to an analysis of variance. Bar charts show the average SRTs for each motion condition. Error bars show standard deviations (linecharts) and 95% CIs (barchart). **B** Average pre-saccadic velocity (PSV) and acceleration (PSA) of this subset. Error bars show 95% CIs.

longer when the saccade target moved toward the current fixation location (foveopetal condition) and shorter when it moved away from it (foveofugal condition).

SRT Asymmetries

Previous studies documented asymmetries in SRTs when saccades moved the eye *away* from a moving stimulus (Tanaka et al., 1998; van Donkelaar, 1999; Khan et al., 2010; Seya and Mori, 2012; Bieg et al., 2013a). Specifically, SRTs in these studies were shorter when the saccade's direction and the stimulus movement direction matched. These asymmetries were explained on the basis of a broad asymmetry of visual-spatial attention in the direction of the movement (e.g., Khan et al., 2010). Could the same argument be used to explain the asymmetries in saccades *to* pursuit that were observed in the current experiment? This seems unlikely since targets stepped to the same hemifield and approximately equal visual field location in both motion direction conditions. Asymmetries in SRTs to moving targets, as they were observed in the current experiment, may therefore not be related to asymmetries in attention.

Saccades *to* moving targets have previously been studied in the context of smooth pursuit eye movements. Suppression of saccades can for example be observed in step-ramp tasks, in which the target is stepped in the opposite movement direction (foveopetal step-ramp). In this case, the target moves across its original position after a specified time, the *zero-crossing* or *eye crossing* time (Gellman and Carl, 1991; de Brouwer et al., 2002a). For zero-crossing times of approximately 200 ms, the initial saccade to the target is suppressed and smooth pursuit of the target commences directly (Rashbass, 1961).

Studies that examined SRTs at the onset of pursuit investigated zero-crossing times near the critical value of approximately 200 ms. For example, Moschner et al. (1999) compared saccades in foveopetal and foveofugal step-ramps. Their reported average SRTs show a substantial difference between both conditions, namely approximately 200 ms in foveofugal and more than twice as long in foveopetal step-ramps. This large difference between motion conditions suggests that the measured saccades in foveopetal trials were mainly catch-up saccades that occurred after zero-crossing. These saccades were planned in the direction of motion after the initial saccade against the motion direction was canceled. Similarly long SRTs were measured by Gellman and Carl (1991) in foveopetal trials with zero-crossing times of 200 ms and 250 ms. In summary, both studies report asymmetries in SRTs that may be primarily related to saccade cancellation.

We recently reported motion-related SRT differences that are not directly related to saccade cancellation and re-planning (Bieg et al., 2013a). In our study, observers were required to alternate their gaze between a central, continuous pursuit task and a peripheral stimulus. For saccades back to pursuit, our results show longer SRTs when the pursuit target moves toward the location of the peripheral stimulus. Critically, these saccades move the eyes against the pursuit motion direction and were initiated before zero-crossing. The current results corroborate and extend this finding. In particular, measurements of SRTs in different step amplitude conditions enabled a comparison of SRTs in eccentricity-matched conditions. This is important to determine whether the SRT differences between the two motion conditions could be explained solely on the basis of different eccentricities of the target at the onset of the saccade (Kalesnykas and Hallett, 1994, but see also Hodgson, 2002; Dafoe et al., 2007). In saccades to moving

94

targets, eccentricity differences are present because of the target's translation during the saccade preparation period. The comparison of SRTs in eccentricity-matched conditions vindicates our main results. This suggests that SRTs to moving targets may not primarily depend on the target's eccentricity but also on the relative motion direction of the saccade target.

Interestingly, the current data also shows a negative correlation between the step amplitude and SRTs in foveopetal trials. Two basic models could explain this observation. First, a general dependency of SRTs on target eccentricity that is modulated by the relative movement direction (fugal/petal) and that is independent of the target's speed. Second, a combination of target eccentricity and speed. The comparison of SRTs between the tested target speed conditions support the last model. We observe clear SRT differences between 20 °/s and 10 °/s eccentricity-matched trials but similar SRTs when comparing conditions with matching zero-crossing times, that is, the combined effect of step size and target speed. Together, this indicates that SRTs of saccades to pursuit increase for zero-crossing times near cancellation (i.e., 200 ms).

A comparison of saccades to pursuit that were matched by saccade amplitudes has been conducted before by Guan et al. (2005). In a study with three primates, similar SRTs were reported in both movement direction conditions. It should however be noted that Guan et al. (2005) tested a single, fairly long (920 ms) zero-crossing time and also focused their analysis on saccade velocity rather than SRT differences.

Finally, it should be mentioned that the obtained increase in SRTs may not only be present in saccades *to* a moving stimulus but also in saccades *from* pursuit. For example, Khan et al. (2010) generally showed that SRTs to targets that were briefly flashed during pursuit were shorter when the target appeared in the direction of pursuit. Considering the retinal signal alone, saccades to a moving target from fixation and saccades during movement to a stationary target are equivalent: The retinal image of the target moves toward the fovea either when the target moves toward the fixation point or when the eye moves toward the target. The data by Khan et al. (2010) could suggest that SRTs are also similarly affected: Their data shows an increase in SRTs if the target was flashed just in front of the pursuit target and thus also in front of the currently fixated location. In this situation, target locations, or rather, the memory of these locations, moves foveopetally during the saccadic latency period. However, it is not clear from this work whether this increase in SRTs can then be attributed to cancellation and re-planning of a saccade, because the eye passed the flashed location for short flash eccentricities, or to a prolonged preparation process in general.

Pre-Saccadic Acceleration

The obtained SRT asymmetries could be related to asymmetries in smooth pursuit initiation (Lisberger and Westbrook, 1985; Tychsen and Lisberger, 1986; Carl and Gellman, 1987). For example, asymmetric pursuit behavior can be observed during the early, open-loop pursuit response that typically occurs before the first saccade to the target. Specifically, targets that move toward the fovea elicit a higher pre-saccadic acceleration (PSA) than targets that move away from the fovea (Lisberger and Westbrook, 1985; Tychsen and Lisberger, 1986). In addition, these studies also report higher accelerations if the target's movement occurs close to the fovea.

The current data is in agreement with these results. PSAs generally occurred in the direction of the pursuit target's motion. No difference in the magnitude between motion conditions could be observed in $10°/s$ trials. In $20°/s$ trials, accelerations were larger in foveopetal trials. In both conditions, accelerations were strongest for trials with small step amplitudes. This is consistent with the observation that pursuit movements toward the fovea exhibit a higher sensitivity to the target's speed and retinal location. For example, the data by Tychsen and Lisberger (1986) show little differences between accelerations at different target speeds for foveofugal target movements but large differences between target speeds for foveopetal trials (see Tychsen and Lisberger, 1986 Fig. 4 A). Similar to the SRT results in foveopetal trials, PSAs are therefore best predicted by taking into account both the step-size and target speed.

The pre-saccadic response typically moves the eye against the upcoming saccade's motion direction in foveopetal trials. Can this kinematic incompatibility explain the asymmetries in SRTs? Our analysis provides evidence against this assumption: asymmetries in

SRTs exist also in a subset of data in which pre-saccadic velocities and accelerations are minimal. This suggests that pre-saccadic behavior may not be a precondition for the occurrence of SRT asymmetries, yet, the similar modulation of both behaviors by step-size and target speed points to a common underlying mechanism that reciprocally affects saccade as well as smooth pursuit initiation.

Neurophysiological Mechanisms

The omnipause neurons (OPNs) in the brainstem regulate both saccadic as well as pursuit eye movements. Inhibition of these neurons is required to trigger a saccade (Scudder et al., 2002) but their activity is also associated with the gain of smooth pursuit eye movements (Missal and Keller, 2002; Kornylo et al., 2003; Krauzlis, 2005). Can this dual function explain the observed effects on SRTs and pre-saccadic accelerations? OPN activity is correlated negatively with the strength of pursuit: Lower OPN activity can be measured during strong pursuit responses (Missal and Keller, 2002). If the level of OPN activity is interpreted as a saccadic threshold (Walton and Gandhi, 2006), this could suggest that saccade initiation is facilitated in conditions with stronger pursuit. The current results speak against this hypothesis: Stronger pre-saccadic pursuit was measured during foveopetal target movements. Considering the ideas above, lower OPN activity and thus shorter SRTs would be expected. Instead, the current SRT results show prolonged SRTs in this condition. This suggests that an interaction between pursuit and saccade that explains the differences in SRTs does not solely occur at the level of OPNs.

Yet, a modulation of OPN activity by cortico-tectal connections can partially explain the results observed in the current and in other studies (e.g., Rashbass, 1961; de Brouwer et al., 2002a). For example, a recent model by Grossberg et al. (2012) assumes increased activation of the rostral superior colliculus (SC) through connections from the middle temporal cortical area (area MT). The rostral SC is associated with maintenance of fixation and excites the OPNs in the brainstem (Munoz and Wurtz, 1993; Hafed et al., 2009). Apart from their general importance for motion perception, neurons in area MT together with those in the middle superior temporal area (MST) provide the signal that drives smooth pursuit eye movements (Newsome et al., 1985; Ilg, 2008). Critically, the model by Grossberg et al. (2012) assumes stronger activation of rostral SC for targets that fall near the foveal area in the MT neuron topography. This implements a foveal threshold that can explain the suppression of a saccade at the onset of pursuit during a foveopetal step-ramp. However, it cannot explain the increased SRTs that were observed in the current study in this condition. A second class of neurons can be found in caudal parts of the SC that are active before and during saccades (Dorris et al., 1997; Dorris and Munoz, 1998; Paré and Hanes, 2003). These neurons are thought to implement a saccadic drive signal whose modulation by excitatory and inhibitory inputs determines SRTs (Wurtz and Goldberg, 1972; Dorris and Munoz, 1998). Fixation neurons generate an intra-collicular signal, which represents one inhibitory input to these saccade-related cells (Munoz and Istvan, 1998; Paré and Hanes, 2003). The activity of fixation neurons can therefore not only shut down saccades entirely but may also influence their onset time.

While the local inhibitory connections within the SC could explain the modulation of SRTs, they cannot explain the asymmetry in SRTs between foveopetal and foveofugal movements that were observed in the current experiment. The observation of asymmetries in pre-saccadic accelerations may lead to a more complete explanation. These accelerations are thought to directly reflect the output of motion processing areas such as MT (Lisberger and Westbrook, 1985; Tychsen and Lisberger, 1986). Investigations of pre-saccadic pursuit suggest a much stronger and speed-dependent activity during foveopetal movements (e.g., Tychsen and Lisberger, 1986 Figs. 2 A and 4 A). Further support for asymmetries in MT comes from psychophysical studies on motion perception, which showed earlier detection of foveopetal motion (Mateeff and Hohnsbein, 1988; Mateeff et al., 1991b; Raymond, 1994; Jancke et al., 2004), and brain imaging studies, which showed higher activation of MT during foveopetal motion (Naito et al., 2000, 2010). Along the lines of the model proposed by Grossberg et al. (2012), stronger MT signals subsequently lead to greater SC fixation neuron activity and thus a depression of saccade-related activity in caudal SC. This depression can lead to prolonged saccadic latencies, in correspondence with the current results.

Fixation and saccade-related activity is not only present in the SC but also more upstream in the frontal eye fields (FEF). Neurons in this area project to the SCs fixation and movement neurons. FEF neuron signals are similar to those in SC in their relationship to saccade initiation and suppression (Hanes and Schall, 1995; Schall et al., 1995; Hanes and Schall, 1996; Hanes et al., 1998). For example, the activity of FEF fixation and saccade-related neurons plays a role in response inhibition or countermanding tasks, which require participants to withhold a saccade once a stop cue is presented (reviewed by Schall and Godlove, 2012; Verbruggen and Logan, 2008). Importantly, recent findings suggest that saccade-related activity in FEF is also influenced by motion information, for example, when a saccadic movement needs to intercept a moving target (Cassanello et al., 2008). This could suggest an alternative pathway by which the saccadic drive signal is modulated, explaining the suppression of saccades at the onset of pursuit and the differences in SRTs that were observed in the current study.

ACKNOWLEDGEMENTS

We thank Sarah Henze for her assistance in data collection. This research was supported by the Max Planck Society, by the myCopter project, funded by the European Commission under the 7th Framework Program, and by the WCU (World Class University) program funded by the Ministry of Education, Science and Technology through the National Research Foundation of Korea (R31-10008).

BIBLIOGRAPHY

Abrams, R. A., Meyer, D. E., and Kornblum, S. (1990). Eye-hand coordination: Oculomotor control in rapid aimed limb movements. *Journal of Experimental Psychology: Human Perception and Performance*, 16(2):248–267.

Andersen, R. A., Brotchie, P. R., and Mazzoni, P. (1992). Evidence for the lateral intraparietal area as the parietal eye field. *Current Opinion in Neurobiology*, 2(6):840–6.

Andersen, R. A., Snyder, L. H., Bradley, D. C., and Xing, J. (1997). Multimodal representation of space in the posterior parietal cortex and its use in planning movements. *Annual Review of Neuroscience*, 20:303–330.

Angelaki, D. E. and Cullen, K. E. (2008). Vestibular system: the many facets of a multimodal sense. *Annual Review of Neuroscience*, 31:125–50.

Angelaki, D. E. and Hess, B. J. M. (2005). Self-motion-induced eye movements: effects on visual acuity and navigation. *Nature Reviews Neuroscience*, 6(12):966–76.

Bach, M. (2007). The Freiburg Visual Acuity Test -Variability unchanged by post-hoc re-analysis. *Graefe's Archive for Clinical and Experimental Ophthalmology*, 245(7):965–71.

Baguley, T. (2012). Calculating and graphing within-subject confidence intervals for ANOVA. *Behavior Research Methods*, 44(1):158–75.

Bahill, A. T., Clark, M. R., and Stark, L. (1975). The Main Sequence, A Tool for Studying Human Eye Movements. *Mathematical Biosciences*, 24:191–204.

Bahill, A. T. and McDonald, J. D. (1983). Smooth pursuit eye movements in response to predictable target motions. *Vision Research*, 23(12):1573–83.

Ball, K. and Sekuler, R. (1980). Human vision favors centrifugal motion. *Perception*, 9(3):317–25.

Baloh, R. W., Sills, A. W., Kumley, W. E., and Hornrubia, V. (1975). Quantitative measurement of saccade amplitude, duration, and velocity. *Neurology*, 25:1065–1070.

Barnes, G. R. (2008). Cognitive processes involved in smooth pursuit eye movements. *Brain and Cognition*, 68(3):309–26.

Barnes, G. R. (2011). Ocular pursuit movements. In Simon, L., Gilchrist, I., and Everling, S., editors, *The Oxford Handbook of Eye Movements*. Oxford University Press, Oxford, UK.

Becker, W. and Fuchs, A. F. (1985). Prediction in the oculomotor system: smooth pursuit during transient disappearance of a visual target. *Experimental Brain Research*, 57(3):562–575.

Becker, W. and Jürgens, R. (1979). An analysis of the saccadic system by means of double step stimuli. *Vision Research*, 19(9):967–83.

Bekkering, H., Adam, J. J., Kingma, H., Huson, A., and Whiting, H. T. A. (1994). Reaction time latencies of eye and hand movements in single- and dual-task conditions. *Experimental Brain Research*, 97(3):471–476.

Bennett, S. J., Orban de Xivry, J.-J., Barnes, G. R., and Lefèvre, P. (2007). Target acceleration can be extracted and represented within the predictive drive to ocular pursuit. *Journal of Neurophysiology*, 98(3):1405–14.

Berryhill, M. E., Chiu, T., and Hughes, H. C. (2006). Smooth pursuit of nonvisual motion. *Journal of Neurophysiology*, 96(1):461–5.

Bieg, H.-J. (2009). Gaze-augmented manual interaction. In *Proceedings of the ACM SIGCHI Conference on Human Factors in Computing Systems (CHI) Doctoral Consortium*, pages 3121–3124, New York, NY, USA. ACM Press.

Bieg, H.-J., Bresciani, J.-P., Bülthoff, H. H., and Chuang, L. L. (2012). Looking for Discriminating is Different from Looking for Lookings Sake. *PLoS ONE*, 7(9):e45445.

Bieg, H.-J., Bresciani, J.-P., Bülthoff, H. H., and Chuang, L. L. (2013a). Saccade reaction time asymmetries during task-switching in pursuit tracking. *Experimental Brain Research*, 230(3):271–281.

Bieg, H.-J., Chuang, L. L., Bülthoff, H. H., and Bresciani, J.-P. (2013b). Asymmetries in saccade reaction times to pursuit (in preparation).

Binsted, G., Chua, R., Helsen, W., and Elliott, D. (2001). Eye-hand coordination in goal-directed aiming. *Human Movement Science*, 20(4-5):563–585.

Bisley, J. W. and Goldberg, M. E. (2010). Attention, intention, and priority in the parietal lobe. *Annual Review of Neuroscience*, 33:1–21.

Bisley, J. W., Mirpour, K., Arcizet, F., and Ong, W. S. (2011). The role of the lateral intraparietal area in orienting attention and its implications for visual search. *The European Journal of Neuroscience*, 33(11):1982–90.

Blohm, G., Missal, M., and Lefèvre, P. (2005). Processing of retinal and extraretinal signals for memory-guided saccades during smooth pursuit. *Journal of Neurophysiology*, 93(3):1510–1522.

Born, R. T. and Bradley, D. C. (2005). Structure and function of visual area MT. *Annual Review of Neuroscience*, 28:157–89.

Boucher, L., Palmeri, T. J., Logan, G. D., and Schall, J. D. (2007). Inhibitory control in mind and brain: an interactive race model of countermanding saccades. *Psychological Review*, 114(2):376–97.

Brefczynski, J. A. and DeYoe, E. A. (1999). A physiological correlate of the 'spotlight' of visual attention. *Nature Neuroscience*, 2(4):370–4.

Bruce, C. J., Goldberg, M. E., Bushnell, M. C., and Stanton, G. B. (1985). Primate frontal eye fields. II. Physiological and anatomical correlates of electrically evoked eye movements. *Journal of Neurophysiology*, 54(3):714–34.

Bruyn, B. D. and Orban, G. (1988). Human velocity and direction discrimination measured with random dot patterns. *Vision Research*, 28(12):1323–1335.

Büttner, U. and Kremmyda, O. (2007). Smooth pursuit eye movements and optokinetic nystagmus. In Straube, A. and Büttner, U., editors, *Neuro-Ophthalmology*. karger, Basel, Switzerland.

Cannon, S. C. and Robinson, D. A. (1987). Loss of the neural integrator of the oculomotor system from brain stem lesions in monkey. *Journal of Neurophysiology*, 57(5):1383–409.

Carl, J. R. and Gellman, R. S. (1987). Human smooth pursuit: stimulus-dependent responses. *Journal of Neurophysiology*, 57(5):1446–63.

Carpenter, R. H. S. (1988). *Movements of the Eyes*. Pion Limited, London, UK, 2nd edition.

Carpenter, R. H. S. and Williams, M. (1995). Neural computation of log likelihood in control of saccadic eye movements. *Nature*, 377(6544):59–62.

Carrasco, M., Penpeci-Talgar, C., and Eckstein, M. P. (2000). Spatial covert attention increases contrast sensitivity across the CSF: support for signal enhancement. *Vision Research*, 40(10-12):1203–15.

Case, G. and Ferrera, V. (2007). Coordination of smooth pursuit and saccade target selection in monkeys. *Journal of Neurophysiology*, 98(4):2206–2214.

Cassanello, C., Nihalani, A., and Ferrera, V. (2008). Neuronal responses to moving targets in monkey frontal eye fields. *Journal of Neurophysiology*, 100(3):1544–1556.

Cattell, J. M. (1886). The influence of the intensity of the stimulus on the length of the reaction time. *Brain*, 8(4):433–576.

Cavegn, D. and Biscaldi, M. (1996). Fixation and saccade control in an express-saccade maker. *Experimental Brain Research*, 109(1):101–16.

Chen-Harris, H., Joiner, W. M., Ethier, V., Zee, D. S., and Shadmehr, R. (2008). Adaptive control of saccades via internal feedback. *The Journal of Neuroscience*, 28(11):2804–13.

Collins, C. E., Lyon, D. C., and Kaas, J. H. (2005). Distribution across cortical areas of neurons projecting to the superior colliculus in new world monkeys. *The Anatomical Record. Part A, Discoveries in Molecular, Cellular, and Evolutionary Biology.*, 285(1):619–27.

Cornsweet, T. N. (1970). *Visual Perception*. Academic Press, New York, New York, USA.

Cynader, M. and Berman, N. (1972). Receptive-field organization of monkey superior colliculus. *Journal of Neurophysiology*, 35(2):187–201.

Dafoe, J. M., Armstrong, I. T., and Munoz, D. P. (2007). The influence of stimulus direction and eccentricity on pro- and anti-saccades in humans. *Experimental Brain Research*, 179(4):563–70.

De Brabander, B., Declerck, C. H., and Boone, C. (2002). Tonic and phasic activation and arousal effects as a function of feedback in repetitive-choice reaction time tasks. *Behavioral Neuroscience*, 116(3):397–402.

de Brouwer, S., Missal, M., Barnes, G., and Lefèvre, P. (2002a). Quantiative analysis of catch-up saccades during sustained pursuit. *Journal of Neurophysiology*, 87:1772–1780.

de Brouwer, S., Yuksel, D., Blohm, G., and Missal, M. (2002b). What Triggers Catch-Up Saccades During Visual Tracking? *Journal of Neurophysiology*, 87:1646–1650.

Deci, E. and Ryan, R. (2000). The "What" and "Why" of goal pursuits: human needs and the self-determination of behavior. *Psychological Inquiry*, 11(4):227–268.

Deno, D., Crandall, W., Sherman, K., and Keller, E. (1995). Characterization of prediction in the primate visual smooth pursuit system. *Biosystems*, 34(94):107–128.

Deubel, H. and Schneider, W. X. (1996). Saccade target selection and object recognition: evidence for a common attentional mechanism. *Vision Research*, 36(12):1827–37.

Dickinson, A. and Balleine, B. (1994). Motivational control of goal-directed action. *Animal Learning & Behavior*, 22(1):1–18.

Dodge, R. (1900). Visual perception during eye movement. *Psychological Review*, 7(5):454–465.

Dorris, M. C. and Munoz, D. P. (1998). Saccadic probability influences motor preparation signals and time to saccadic initiation. *The Journal of Neuroscience*, 18(17):7015–26.

Dorris, M. C., Paré, M., and Munoz, D. P. (1997). Neuronal activity in monkey superior colliculus related to the initiation of saccadic eye movements. *The Journal of Neuroscience*, 17(21):8566–79.

Duchowski, A. T. (2007). *Eye Tracking Methodology*. Springer-Verlag, London, UK, 2nd edition.

Dukelow, S. P., DeSouza, J. F., Culham, J. C., van den Berg, A. V., Menon, R. S., and Vilis, T. (2001). Distinguishing subregions of the human MT+ complex using visual fields and pursuit eye movements. *Journal of Neurophysiology*, 86(4):1991–2000.

Duvernoy, H. M. (2007). *The Human Brain*. Springer-Verlag, Wien, 2nd edition.

Eggert, T., Guan, Y., Bayer, O., and Büttner, U. (2005a). Saccades to moving targets. *Annals of the New York Academy of Sciences*, 1039:149–59.

Eggert, T., Rivas, F., and Straube, A. (2005b). Predictive strategies in interception tasks: differences between eye and hand movements. *Experimental Brain Research*, 160(4):433–49.

Epelboim, J., Steinman, R. M., Kowler, E., Pizlo, Z., Erkelens, C. J., and Collewijn, H. (1997). Gaze-shift dynamics in two kinds of sequential looking tasks. *Vision Research*, 37(18):2597–2607.

Etchells, P. and Benton, C. (2010). The target velocity integration function for saccades. *Journal of Vision*, 10(6):1–14.

Fecteau, J. H. and Munoz, D. P. (2006). Salience, relevance, and firing: a priority map for target selection. *Trends in Cognitive Sciences*, 10(8):382–90.

Ferrera, V. P. and Lisberger, S. G. (1995). Attention and Target Selection for Smooth Pursuit Eye Movements. *The Journal of Neuroscience*, 15(11):7472–7484.

Findlay, J. M. and Gilchrist, I. D. (2003). *Active Vision: The psychology of looking and seeing*. Oxford University Press, Oxford, UK.

Findlay, J. M. and Walker, R. (1999). A model of saccade generation based on parallel processing and competitive inhibition. *Behavioral and Brain Sciences*, 22(4):661–74; discussion 674–721.

Fischer, B. and Boch, R. (1983). Saccadic eye movements after extremely short reaction times in the monkey. *Brain Research*, 260(1):21–26.

Fischer, B., Weber, H., Biscaldi, M., Aiple, F., Otto, P., and Stuhr, V. (1993). Separate populations of visually guided saccades in humans: reaction times and amplitudes. *Experimental Brain Research*, 92(3):528–541.

Freedman, E. G., Stanford, T. R., and Sparks, D. L. (1996). Combined eye-head gaze shifts produced by electrical stimulation of the superior colliculus in rhesus monkeys. *Journal of Neurophysiology*, 76(2):927–952.

Frens, M. A. and van der Geest, J. N. (2002). Scleral search coils influence saccade dynamics. *Journal of Neurophysiology*, 88(2):692–8.

Freyberg, S. and Ilg, U. J. (2008). Anticipatory smooth-pursuit eye movements in man and monkey. *Experimental Brain Research*, 186(2):203–14.

Fries, W. (1984). Cortical projections to the superior colliculus in the macaque monkey: a retrograde study using horseradish peroxidase. *The Journal of Comparative Neurology*, 230(1):55–76.

Fuchs, A. and Luschei, E. (1970). Firing patterns of abducens neurons of alert monkeys in relationship to horizontal eye movement. *Journal of Neurophysiology*, 33(3):382–392.

Fuchs, A. F. (1967). Saccadic and smooth pursuit eye movements in the monkey. *The Journal of Physiology*, 191(3):609–631.

Fuchs, A. F. and Binder, M. D. (1983). Fatigue resistance of human extraocular muscles. *Journal of Neurophysiology*, 49(1):28–34.

Fuller, J. H. (1996). Eye position and target amplitude effects on human visual saccadic latencies. *Experimental Brain Research*, 109(3):457–66.

Gandhi, N. J. and Katnani, H. a. (2011). Motor functions of the superior colliculus. *Annual Review of Neuroscience*, 34:205–31.

Garbutt, S., Harwood, M. R., and Harris, C. M. (2001). Comparison of the main sequence of reflexive saccades and the quick phases of optokinetic nystagmus. *The British Journal of Ophthalmology*, 85(12):1477–83.

Gardner, J. and Lisberger, S. (2001). Linked target selection for saccadic and smooth pursuit eye movements. *The Journal of Neuroscience*, 21(6):2075–2084.

Gauthier, G. M. and Hofferer, J. (1976). Eye Tracking of Self-Moved Targets in the Absence of Vision. *Experimental Brain Research*, 21(2):121–139.

Gauthier, G. M., Vercher, J.-L., Mussa Ivaldi, F., and Marchetti, E. (1988). Oculomanual tracking of visual targets: control learning, coordination control and coordination model. *Experimental Brain Research*, 73(1):127–137.

Gellman, R. S. and Carl, J. R. (1991). Motion processing for saccadic eye movements in humans. *Experimental Brain Research*, 84(3):660–667.

Gilchrist, I. D. (2011). Saccades. In Liversedge, S. P., Gilchrist, I., and Everling, S., editors, *The Oxford Handbook of Eye Movements*. Oxford University Press, Oxford, UK.

Girard, B. and Berthoz, A. (2005). From brainstem to cortex: computational models of saccade generation circuitry. *Progress in Neurobiology*, 77(4):215–51.

Glimcher, P. W. (2003). The neurobiology of visual-saccadic decision making. *Annual Review of Neuroscience*, 26:133–79.

Godijn, R. and Theeuwes, J. (2003). The Relationship Between Exogenous and Endogenous Saccades and Attention. In Hyönä, J., Radach, R., and Deubel, H., editors, *The Minds Eye: Cognitive and Applied Aspects of Eye Movement Research*. North Holland, Amsterdam, Netherlands.

Gold, J. I. and Shadlen, M. N. (2007). The Neural Basis of Decision Making. *Annual Review of Neuroscience*, 30:535–574.

Goldberg, J. H. and Wichansky, A. M. (2003). Eye Tracking in Usability Evaluation: A Practitioner's Guide. In Hyönä, J., Radach, R., and Deubel, H., editors, *The Mind's Eye: Cognitive and Applied Aspects of Eye Movement Research*. North Holland, Amsterdam, Netherlands.

Goossens, H. H. L. M. and Van Opstal, A. J. (2006). Dynamic ensemble coding of saccades in the monkey superior colliculus. *Journal of Neurophysiology*, 95(4):2326–41.

Gottesman, I. and Gould, T. (2003). The endophenotype concept in psychiatry: etymology and strategic intentions. *American Journal of Psychiatry*, (April):636–645.

Grossberg, S., Srihasam, K., and Bullock, D. (2012). Neural dynamics of saccadic and smooth pursuit eye movement coordination during visual tracking of unpredictably moving targets. *Neural Networks*, 27:1–20.

Guan, Y., Eggert, T., Bayer, O., and Büttner, U. (2005). Saccades to stationary and moving targets differ in the monkey. *Experimental Brain Research*, 161(2):220–32.

Guyader, N., Malsert, J., and Marendaz, C. (2010). Having to identify a target reduces latencies in prosaccades but not in antisaccades. *Psychological Research*, 74(1):12–20.

Hafed, Z. M., Goffart, L., and Krauzlis, R. J. (2009). A neural mechanism for microsaccade generation in the primate superior colliculus. *Science*, 323(5916):940–3.

Hanes, D. and Wurtz, R. (2001). Interaction of the frontal eye field and superior colliculus for saccade generation. *Journal of Neurophysiology*, 85(2):804–815.

Hanes, D. P., Patterson, W. F., and Schall, J. D. (1998). Role of frontal eye fields in countermanding saccades: visual, movement, and fixation activity. *Journal of Neurophysiology*, 79(2):817–34.

Hanes, D. P. and Schall, J. D. (1995). Countermanding saccades in macaque. *Visual Neuroscience*, 12(5):929–37.

Hanes, D. P. and Schall, J. D. (1996). Neural control of voluntary movement initiation. *Science*, 274(5286):427–30.

Hanes, D. P., Thompson, K. G., and Schall, J. D. (1995). Relationship of presaccadic activity in frontal eye field and supplementary eye field to saccade initiation in macaque: Poisson spike train analysis. *Experimental Brain Research*, 103(1):85–96.

Harris, C. M. and Wolpert, D. M. (1998). Signal-dependent noise determines motor planning. *Nature*, 394(6695):780–784.

Harris, C. M. and Wolpert, D. M. (2006). The main sequence of saccades optimizes speed-accuracy trade-off. *Biological Cybernetics*, 95(1):21–9.

Heinen, S. J., Jin, Z., and Watamaniuk, S. N. J. (2011). Flexibility of foveal attention during ocular pursuit. *Journal of Vision*, 11:1–12.

Heywood, S. and Churcher, J. (1981). Saccades to step-ramp stimuli. *Vision Research*, 21(4):479–90.

Hikosaka, O., Takikawa, Y., and Kawagoe, R. (2000). Role of the basal ganglia in the control of purposive saccadic eye movements. *Physiological Reviews*, 80(3):953–78.

Hodgson, T. L. (2002). The location marker effect. Saccadic latency increases with target eccentricity. *Experimental Brain Research*, 145(4):539–42.

Hoffman, J. E. and Subramaniam, B. (1995). The role of visual attention in saccadic eye movements. *Perception & Psychophysics*, 57(6):787–95.

Ilg, U. J. (1997). Slow Eye Movements. *Progress in Neurobiology*, 53:293–329.

Ilg, U. J. (2008). The role of areas MT and MST in coding of visual motion underlying the execution of smooth pursuit. *Vision Research*, 48(20):2062–9.

Ilg, U. J. and Thier, P. (2003). Visual tracking neurons in primate area MST are activated by smooth-pursuit eye movements of an "imaginary" target. *Journal of Neurophysiology*, 90(3):1489–502.

Itti, L. and Koch, C. (2001). Computational modelling of visual attention. *Nature Reviews Neuroscience*, 2(3):194–203.

Jacob, R. J. K. (1991). The Use of Eye Movements in Interaction Techniques: What You Look At is What You Get. *ACM Transactions on Information Systems*, 9(3):152–169.

Jacob, R. J. K. and Kam, K. S. (2003). Eye Tracking in Human-Computer Interaction and Usability Research: Ready to Deliver the Promises. In Hyönä, J., Radach, R., and Deubel, H., editors, *The Mind's Eye: Cognitive and Applied Aspects of Eye Movement Research*, pages 573–605. North Holland, Amsterdam, Netherlands.

Jancke, D., Erlhagen, W., Schöner, G., and Dinse, H. R. (2004). Shorter latencies for motion trajectories than for flashes in population responses of cat primary visual cortex. *The Journal of Physiology*, 556(Pt 3):971–82.

Jantz, J. J., Watanabe, M., Everling, S., and Munoz, D. P. (2013). Threshold mechanism for saccade initiation in frontal eye field and superior colliculus. *Journal of Neurophysiology*, 109(11):2767–80.

Javal, E. (1879). Essai sur la physiologie de la lecture. *Annales D'Oculistique*, 82:242–253.

Jonikaitis, D., Deubel, H., and De'Sperati, C. (2009). Time gaps in mental imagery introduced by competing saccadic tasks. *Vision Research*, 49(17):2164–75.

Jürgens, R., Becker, W., and Kornhuber, H. H. (1981). Natural and drug-induced variations of velocity and duration of human saccadic eye movements: evidence for a control of the neural pulse generator by local feedback. *Biological Cybernetics*, 39(2):87–96.

Kalesnykas, R. and Hallett, P. E. (1994). Retinal eccentricity and the latency of eye saccades. *Vision Research*, 34(4):517–531.

Kanai, R., van der Geest, J. N., and Frens, M. A. (2003). Inhibition of saccade initiation by preceding smooth pursuit. *Experimental Brain Research*, 148(3):300–7.

Kao, G. W. and Morrow, M. J. (1994). The relationship of anticipatory smooth eye movement to smooth pursuit initiation. *Vision Research*, 34(22):3027–36.

Kathmann, N., Hochrein, A., and Uwer, R. (1999). Effects of dual task demands on the accuracy of smooth pursuit eye movements. *Psychophysiology*, 36(2):158–63.

Kawagoe, R., Takikawa, Y., and Hikosaka, O. (1998). Expectation of reward modulates cognitive signals in the basal ganglia. *Nature Neuroscience*, 1(5):411–416.

Keller, E. and Johnsen, S. D. (1990). Velocity prediction in corrective saccades during smooth-pursuit eye movements in monkey. *Experimental Brain Research*, 80(3):525–31.

Khan, A., Lefèvre, P., Heinen, S., and Blohm, G. (2010). The default allocation of attention is broadly ahead of smooth pursuit. *Journal of Vision*, 10(13):1–17.

Kim, C. E., Thaker, G. K., Ross, D. E., and Medoff, D. (1997). Accuracies of saccades to moving targets during pursuit initiation and maintenance. *Experimental Brain Research*, 113(2):371–7.

Kline, R. B. (2005). *Beyond Significance Testing*. American Psychological Association, Washington, DC, USA.

Koken, P. W. and Erkelens, C. J. (1992). Influences of hand movements on eye movements in tracking tasks in man. *Experimental Brain Research*, 88(3):657–664.

Kornylo, K., Dill, N., Saenz, M., and Krauzlis, R. J. (2003). Cancelling of pursuit and saccadic eye movements in humans and monkeys. *Journal of Neurophysiology*, 89(6):2984–99.

Kowler, E., Anderson, E., Dosher, B., and Blaser, E. (1995). The role of attention in the programming of saccades. *Vision Research*, 35(13):1897–916.

Krauzlis, R. (2003). Neuronal activity in the rostral superior colliculus related to the initiation of pursuit and saccadic eye movements. *The Journal of Neuroscience*, 23(10):4333–4344.

Krauzlis, R. J. (2005). The control of voluntary eye movements: new perspectives. *The Neuroscientist*, 11(2):124–37.

Krauzlis, R. J. and Miles, F. A. (1996). Decreases in the latency of smooth pursuit and saccadic eye movements produced by the "gap paradigm" in the monkey. *Vision Research*, 36(13):1973–85.

Land, M. F. (2009). Vision, eye movements, and natural behavior. *Visual Neuroscience*, 26(1):51–62.

Land, M. F. and Hayhoe, M. M. (2001). In what ways do eye movements contribute to everyday activities? *Vision Research*, 41(25-26):3559–65.

Land, M. F., Mennie, N., and Rusted, J. (1999). The roles of vision and eye movements in the control of activities of daily living. *Perception*, 28(11):1311–1328.

Land, M. F. and Tatler, B. W. (2009). *Looking and Acting - Vision and Eye Movements in Natural Behaviour*. Oxford University Press, Oxford, UK.

Larson, A. and Loschky, L. (2009). The contributions of central versus peripheral vision to scene gist recognition. *Journal of Vision*, 9(10):1–16.

Lauwereyns, J., Watanabe, K., Coe, B., and Hikosaka, O. (2002). A neural correlate of response bias in monkey caudate nucleus. *Nature*, 418(6896):413–417.

Lebedev, S., Van Gelder, P., and Tsui, W. H. (1996). Square-root relations between main saccadic parameters. *Investigative Ophthalmology & Visual Science*, 37(13):2750–8.

Leigh, R. J. and Kennard, C. (2004). Using saccades as a research tool in the clinical neurosciences. *Brain*, 127(Pt 3):460–77.

Leigh, R. J. and Zee, D. S. (2006). *The Neurology of Eye Movements*. Oxford University Press, Oxford, UK.

Levy, D. L., Holzman, P. S., Matthysse, S., and Mendell, N. R. (1993). Eye tracking dysfunction and schizophrenia: a critical perspective. *Schizophrenia Bulletin*, 19(3):461–536.

Lindner, A. and Ilg, U. J. (2006). Suppression of optokinesis during smooth pursuit eye movements revisited: the role of extra-retinal information. *Vision Research*, 46(6-7):761–7.

Lisberger, S. and Westbrook, L. (1985). Properties of visual inputs that initiate horizontal smooth pursuit eye movements in monkeys. *The Journal of Neuroscience*, 5(6):1662–1673.

Lisberger, S. G. (1998). Postsaccadic enhancement of initiation of smooth pursuit eye movements in monkeys. *Journal of Neurophysiology*, 79(4):1918–30.

Lisberger, S. G., Morris, E. J., and Tychsen, L. (1987). Visual motion processing and sensory-motor integration for smooth pursuit eye movements. *Annual Review of Neuroscience*, 10:97–129.

Liversedge, S. P. and Findlay, J. M. (2000). Saccadic eye movements and cognition. *Trends in Cognitive Sciences*, 4(1):6–14.

Logan, G. D., Cowan, W. B., and Davis, K. A. (1984). On the ability to inhibit simple and choice reaction time responses: a model and a method. *Journal of Experimental Psychology: Human Perception and Performance*, 10(2):276–91.

Lovejoy, L. P., Fowler, G. a., and Krauzlis, R. J. (2009). Spatial allocation of attention during smooth pursuit eye movements. *Vision Research*, 49(10):1275–1285.

Lünenburger, L. and Hoffmann, K.-P. (2003). Arm movement and gap as factors influencing the reaction time of the second saccade in a double-step task. *European Journal of Neuroscience*, 17(11):2481–2491.

Lünenburger, L., Kutz, D. F., and Hoffmann, K.-P. (2000). Influence of arm movements on saccades in humans. *European Journal of Neuroscience*, 12(11):4107–4116.

Madelain, L., Paeye, C., and Darcheville, J.-C. (2011a). Operant control of human eye movements. *Behavioural Processes*, 87(1):142–8.

Madelain, L., Paeye, C., and Wallman, J. (2011b). Modification of saccadic gain by reinforcement. *Journal of Neurophysiology*, 106(1):219.

Mansfield, R. J. W. (1973). Latency Functions in Human Vision. *Vision Research*, 13:2219–2234.

Martinez-Conde, S., Macknik, S. L., and Hubel, D. H. (2004). The role of fixational eye movements in visual perception. *Nature Reviews Neuroscience*, 5(3):229–40.

Mateeff, S., Bohdanecky, Z., Hohnsbein, J., Ehrenstein, W. H., and Yakimoff, N. (1991a). A constant latency difference determines directional anisotropy in visual motion perception. *Vision Research,* 31(12):2235–7.

Mateeff, S. and Hohnsbein, J. (1988). Perceptual latencies are shorter for motion towards the fovea than for motion away. *Vision Research,* 28(6):711–9.

Mateeff, S., Yakimoff, N., and Hohnsbein, J. (1991b). Selective directional sensitivity in visual motion perception. *Vision Research,* 31(1):131–138.

Mather, J. A. and Putchat, C. (1983). Parallel ocular and manual tracking responses to a continuously moving visual target. *Journal of Motor Behavior,* 15(1):29–38.

McDowell, J. E., Clementz, B. A., and Sweeney, J. A. (2011). Eye movements in psychiatric patients. In Simon, L., Gilchrist, I., and Everling, S., editors, *The Oxford Handbook of Eye Movements.* Oxford University Press, Oxford, UK.

McPeek, R. M., Han, J. H., and Keller, E. L. (2003). Competition between saccade goals in the superior colliculus produces saccade curvature. *Journal of Neurophysiology,* 89(5):2577–90.

Meyer, C., Lasker, A., and Robinson, D. A. (1985). The upper limit of human smooth pursuit velocity. *Vision Research,* 25(1):561–563.

Milstein, D. M. and Dorris, M. C. (2007). The influence of expected value on saccadic preparation. *The Journal of Neuroscience,* 27(18):4810–4818.

Missal, M. and Keller, E. L. (2002). Common inhibitory mechanism for saccades and smooth-pursuit eye movements. *Journal of Neurophysiology,* 88(4):1880–92.

Montagnini, A. and Chelazzi, L. (2005). The urgency to look: prompt saccades to the benefit of perception. *Vision Research,* 45(27):3391–3401.

Moore, T. and Armstrong, K. (2003). Selective gating of visual signals by microstimulation of frontal cortex. *Nature,* 421(6921):370–373.

Moore, T. and Fallah, M. (2001). Control of eye movements and spatial attention. *Proceedings of the National Academy of Sciences of the United States of America,* 98(3):1273–6.

Morey, R. D. (2008). Confidence intervals from normalized data: a correction to Cousineau (2005). *Tutorial in Quantitative Methods for Psychology,* 4(2):61–64.

Moschner, C., Crawford, T. J., Heide, W., Trillenberg, P., Kömpf, D., and Kennard, C. (1999). Deficits of smooth pursuit initiation in patients with degenerative cerebellar lesions. *Brain*, 122(11):2147–58.

Moschovakis, A. K., Scudder, C. A., and Highstein, S. M. (1996). The microscopic anatomy and physiology of the mammalian saccadic system. *Progress in Neurobiology*, 50(2-3):133–254.

Munoz, D. P. and Everling, S. (2004). Look away: the anti-saccade task and the voluntary control of eye movement. *Nature Reviews Neuroscience*, 5(3):218–28.

Munoz, D. P. and Fecteau, J. H. (2002). Vying for dominance: dynamic interactions control visual fixation and saccadic initiation in the superior colliculus. *Progress in Brain research*, 140:3–19.

Munoz, D. P. and Istvan, P. J. (1998). Lateral inhibitory interactions in the intermediate layers of the monkey superior colliculus. *Journal of Neurophysiology*, 79(3):1193–209.

Munoz, D. P. and Wurtz, R. H. (1993). Fixation cells in monkey superior colliculus. I. Characteristics of cell discharge. *Journal of Neurophysiology*, 70(2):559–575.

Naito, T., Kaneoke, Y., Osaka, N., and Kakigi, R. (2000). Asymmetry of the human visual field in magnetic response to apparent motion. *Brain Research*, 865(2):221–6.

Naito, T., Sato, H., and Osaka, N. (2010). Direction anisotropy of human motion perception depends on stimulus speed. *Vision Research*, 50(18):1862–6.

Navalpakkam, V., Koch, C., Rangel, A., and Perona, P. (2010). Optimal reward harvesting in complex perceptual environments. *Proceedings of the National Academy of Sciences of the United States of America*, 107(11):5232–5237.

Neggers, S. F. W. and Bekkering, H. (2000). Ocular gaze is anchored to the target of an ongoing pointing movement. *Journal of Neurophysiology*, 83:639–651.

Newsome, W. T., Wurtz, R. H., Dürsteler, M. R., and Mikami, A. (1985). Deficits in visual motion processing following ibotenic acid lesions of the middle temporal visual area of the macaque monkey. *The Journal of Neuroscience*, 5(3):825–40.

Oommen, B. S. and Stahl, J. S. (2005). Amplitudes of head movements during putative eye-only saccades. *Brain Research*, 1065(1-2):68–78.

Orban de Xivry, J.-J. and Lefèvre, P. (2007). Saccades and pursuit: two outcomes of a single sensorimotor process. *The Journal of Physiology,* 584(1):11–23.

Paré, M. and Dorris, M. C. (2011). The role of posterior parietal cortex in the regulation of saccadic eye movements. In Liversedge, S. P., Gilchrist, I. D., and Everling, S., editors, *The Oxford Handbook of Eye Movements.* Oxford, UK.

Paré, M. and Hanes, D. P. (2003). Controlled movement processing: superior colliculus activity associated with countermanded saccades. *The Journal of Neuroscience,* 23(16):6480–9.

Paré, M. and Munoz, D. P. (2001). Expression of a re-centering bias in saccade regulation by superior colliculus neurons. *Experimental Brain Research,* 137(3-4):354–368.

Pelisson, D. and Prablanc, C. (1988). Kinematics of centrifugal and centripetal saccadic eye movements in man. *Vision Research,* 28(1):87–94.

Platt, M. L. and Glimcher, P. W. (1999). Neural correlates of decision variables in parietal cortex. *Nature,* 400(6741):233–8.

Posner, M. and Dehaene, S. (1994). Attentional networks. *Trends in Neurosciences,* 17(2):75–79.

Posner, M. I. (1980). Orienting of attention. *The Quarterly Journal of Experimental Psychology,* 32(1):3–25.

Prablanc, C., Echallier, J. F., Komilis, E., and Jeannerod, M. (1979). Optimal response of eye and hand motor systems in pointing at a visual target. I. Spatio-temporal characteristics of eye and hand movements and their relationships when varying the amount of visual information. *Biological Cybernetics,* 35(2):113–124.

Prinzmetal, W., McCool, C., and Park, S. (2005). Attention: reaction time and accuracy reveal different mechanisms. *Journal of Experimental Psychology: General,* 134(1):73–92.

Prsa, M., Dicke, P. W., and Thier, P. (2010). The absence of eye muscle fatigue indicates that the nervous system compensates for non-motor disturbances of oculomotor function. *The Journal of Neuroscience,* 30(47):15834–15842.

Quaia, C., Lefèvre, P., and Optican, L. M. (1999). Model of the control of saccades by superior colliculus and cerebellum. *Journal of Neurophysiology,* 82(2):999–1018.

Rashbass, C. (1961). The relationship between saccadic and smooth tracking eye movements. *The Journal of Physiology*, 159:326–338.

Ratcliff, R. (1993). Methods for dealing with reaction time outliers. *Psychological Bulletin*, 114(3):510–32.

Raymond, J. E. (1994). Directional anisotropy of motion sensitivity across the visual field. *Vision Research*, 34(8):1029–37.

Rayner, K. (2009). Eye movements and attention in reading, scene perception, and visual search. *Quarterly Journal of Experimental Psychology*, 62(8):1457–506.

Reddi, B. A. J., Asrress, K., and Carpenter, R. H. S. (2003). Accuracy, information, and response time in a saccadic decision task. *Journal of Neurophysiology*, 90(5):3538.

Reddi, B. A. J. and Carpenter, R. H. S. (2000). The influence of urgency on decision time. *Nature Neuroscience*, 3(8):827–830.

Reulen, J. P. H. (1984). Latency of Visually Evoked Saccadic Eye Movements. II. Temporal Properties of the Facilitation Mechanism. *Biological Cybernetics*, 50:263–271.

Reuter-Lorenz, P. A., Hughes, H. C., and Fendrich, R. (1991). The reduction of saccadic latency by prior offset of the fixation point: an analysis of the gap effect. *Perception & Psychophysics*, 49(2):167–75.

Rizzolatti, G., Riggio, L., Dascola, I., and Umiltá, C. (1987). Reorienting attention across the horizontal and vertical meridians: evidence in favor of a premotor theory of attention. *Neuropsychologia*, 25(1A):31–40.

Robinson, D. A. (1964). The mechanics of human saccadic eye movement. *The Journal of Physiology*, pages 245–264.

Robinson, D. A. (1972). Eye movements evoked by collicular stimulation in the alert monkey. *Vision Research*, 12(11):1795–808.

Robinson, D. A., Gordon, J. L., and Gordon, S. E. (1986). A Model of the Smooth Pursuit Eye Movement System. *Biological Cybernetics*, 57:43–57.

Ross, J., Goldberg, M. E., Morrone, M. C., and Burr, D. C. (2001). Changes in visual perception at the time of saccades. *Trends in Neurosciences*, 24(2):113–21.

Salmoni, A. W., Schmidt, R. A., and Walter, C. B. (1984). Knowledge of results and motor learning: a review and critical reappraisal. *Psychological Bulletin*, 95(3):355–86.

Salvucci, D. D. and Anderson, J. R. (2000). Intelligent gaze-added interfaces. *Proceedings of the SIGCHI Conference on Human Factors in Computing Systems (CHI)*, pages 273–280.

Sanders, A. F. (1970). Some aspects of the selective process in the functional visual field. *Ergonomics*, 13(1):101–17.

Saslow, M. G. (1967). Effecs of components of displacement-step stimuli upon latency of saccadic eye movement. *Journal of the Optical Society of America*, 57(8):1024–1029.

Schall, J. D. (2004). On building a bridge between brain and behavior. *Annual Review of Psychology*, 55:23–50.

Schall, J. D. and Boucher, L. (2007). Executive control of gaze by the frontal lobes. *Cognitive, Affective & Behavioral Neuroscience*, 7(4):396–412.

Schall, J. D. and Godlove, D. C. (2012). Current advances and pressing problems in studies of stopping. *Current Opinion in Neurobiology*, 22(6):1012–21.

Schall, J. D., Morel, A., King, D. J., and Bullier, J. (1995). Topography of visual cortex connections with frontal eye field in macaque: convergence and segregation of processing streams. *The Journal of Neuroscience*, 15(6):4464–87.

Schall, J. D., Purcell, B. A., Heitz, R. P., Logan, G. D., and Palmeri, T. J. (2011). Neural mechanisms of saccade target selection: gated accumulator model of the visual-motor cascade. *The European Journal of Neuroscience*, 33(11):1991–2002.

Schall, J. D. and Thompson, K. G. (1999). Neural selection and control of visually guided eye movements. *Annual Review of Neuroscience*, 22:241–59.

Schiller, P., True, S. D., and Conway, J. L. (1980). Deficits in eye movements following frontal eye-field and superior colliculus ablations. *Journal of Neurophysiology*, 44(6):1175–1189.

Schmidt, D., Abel, L. A., Dell'Osso, L. F., and Daroff, R. B. (1979). Saccadic velocity characteristics: intrinsic variability and fatigue. *Aviation, Space, and Environmental Medicine*, 50(4):393–5.

Scudder, C. A., Kaneko, C. S., and Fuchs, A. F. (2002). The brainstem burst generator for saccadic eye movements: a modern synthesis. *Experimental Brain Research*, 142(4):439–62.

Segraves, M. and Goldberg, M. (1987). Functional properties of corticotectal neurons in the monkey's frontal eye field. *Journal of Neuro-*

physiology, pages 1387–1419.

Sereno, M. I., Pitzalis, S., and Martinez, A. (2001). Mapping of contralateral space in retinotopic coordinates by a parietal cortical area in humans. *Science*, 294(5545):1350–4.

Serra, A., Derwenskus, J., Downey, D. L., and Leigh, R. J. (2003). Role of eye movement examination and subjective visual vertical in clinical evaluation of multiple sclerosis. *Journal of Neurology*, 250(5):569–75.

Seya, Y. and Mori, S. (2012). Spatial attention and reaction times during smooth pursuit eye movement. *Attention, Perception, & Psychophysics*, 74(3):493–509.

Shadlen, M. N. and Newsome, W. T. (2001). Neural basis of a perceptual decision in the parietal cortex (area LIP) of the rhesus monkey. *Journal of Neurophysiology*, 86(4):1916–36.

Shadmehr, R., Orban De Xivry, J.-J., Xu-Wilson, M., and Shih, T.-Y. (2010). Temporal discounting of reward and the cost of time in motor control. *The Journal of Neuroscience*, 30(31):10507–16.

Shagass, C., Roemer, R., and Amadeo, M. (1976). Eye-tracking performance and engagement of attention. *Archives of General Psychiatry*, 33:121–125.

Smeets, J. B. and Bekkering, H. (2000). Prediction of saccadic amplitude during smooth pursuit eye movements. *Human Movement Science*, 19(3):275–295.

Snyder, L. H., Calton, J. L., Dickinson, A. R., and Lawrence, B. M. (2002). Eye-hand coordination: saccades are faster when accompanied by a coordinated arm movement. *Journal of Neurophysiology*, 87:2279–2286.

Sommer, M. A. and Wurtz, R. H. (2008). Brain circuits for the internal monitoring of movements. *Annual Review of Neuroscience*, 31:317–38.

Sparks, D. L. (2002). The brainstem control of saccadic eye movements. *Nature Reviews Neuroscience*, 3(12):952–64.

Spering, M., Gegenfurtner, K. R., and Kerzel, D. (2006). Distractor interference during smooth pursuit eye movements. *Journal of Experimental Psychology: Human Perception and Performance*, 32(5):1136–54.

Stanton, G. B., Goldberg, M. E., and Bruce, C. J. (1988). Frontal eye field efferents in the macaque monkey: I. Subcortical pathways and topography of striatal and thalamic terminal fields. *The Journal of Comparative Neurology*, 271(4):473–92.

Stark, L., Adler, D., and Bahill, A. T. (1975). Most naturally occurring human saccades have magnitudes of 15 degrees or less. *Investigative Ophthalmology*, 14(6):468–469.

Steenrod, S. C., Phillips, M. H., and Goldberg, M. E. (2013). The lateral intraparietal area codes the location of saccade targets and not the dimension of the saccades that will be made to acquire them. *Journal of Neurophysiology*, 109(10):2596–605.

Straube, A., Fuchs, A. F., Usher, S., and Robinson, F. R. (1997). Characteristics of saccadic gain adaptation in rhesus macaques. *Journal of Neurophysiology*, 77(2):874–95.

Stritzke, M., Trommershäuser, J., and Gegenfurtner, K. R. (2009). Effects of salience and reward information during saccadic decisions under risk. *Journal of the Optical Society of America: A, Optics, Image Science, and Vision*, 26(11):B1–13.

Sweeney, J. A., Clementz, B. A., Haas, G. L., Escobar, M. D., Drake, K., and Frances, A. J. (1994). Eye tracking dysfunction in schizophrenia: characterization of component eye movement abnormalities, diagnostic specificity, and the role of attention. *Journal of Abnormal Psychology*, 103(2):222–30.

Sweeney, J. A., Takarae, Y., Macmillan, C., Luna, B., and Minshew, N. J. (2004). Eye movements in neurodevelopmental disorders. *Current Opinion in Neurology*, 17(1):37.

Takikawa, Y., Kawagoe, R., Itoh, H., Nakahara, H., and Hikosaka, O. (2002). Modulation of saccadic eye movements by predicted reward outcome. *Experimental Brain Research*, 142(2):284–291.

Tanabe, J., Tregellas, J., Miller, D., Ross, R. G., and Freedman, R. (2002). Brain Activation during Smooth-Pursuit Eye Movements. *NeuroImage*, 17(3):1315–1324.

Tanaka, M., Yoshida, T., and Fukushima, K. (1998). Latency of saccades during smooth-pursuit eye movement in man. Directional asymmetries. *Experimental Brain Research*, 121(1):92–8.

Tedeschi, G., Di Costanzo, A., Allocca, S., Toriello, A., Ammendola, A., Quattrone, A., and Bonavita, V. (1991). Saccadic eye movements analysis in the early diagnosis of myasthenia gravis. *Italian Journal of Neurological Sciences*, 12(4):389–95.

Theeuwes, J. and Belopolsky, A. V. (2012). Reward grabs the eye: oculomotor capture by rewarding stimuli. *Vision Research*, 74:80–5.

Theeuwes, J., Kramer, A., Hahn, S., and Irwin, D. E. (1998). Our eyes do not always go where we want them to go: Capture of the eyes by new objects. *Psychological Science*, 9(5):379.

Thier, P. and Ilg, U. J. (2005). The neural basis of smooth-pursuit eye movements. *Current Opinion in Neurobiology*, 15(6):645–52.

Tootell, R. B., Reppas, J. B., Kwong, K. K., Malach, R., Born, R. T., Brady, T. J., Rosen, B. R., and Belliveau, J. W. (1995). Functional analysis of human MT and related visual cortical areas using magnetic resonance imaging. *The Journal of Neuroscience*, 15(4):3215–30.

Träisk, F., Bolzani, R., and Ygge, J. (2005). A comparison between the magnetic scleral search coil and infrared reflection methods for saccadic eye movement analysis. *Graefe's archive for Clinical and Experimental Ophthalmology*, 243(8):791–7.

Trappenberg, T. P., Dorris, M. C., Munoz, D. P., and Klein, R. M. (2001). A model of saccade initiation based on the competitive integration of exogenous and endogenous signals in the superior colliculus. *Journal of Cognitive Neuroscience*, 13(2):256–71.

Trillenberg, P., Lencer, R., and Heide, W. (2004). Eye movements and psychiatric disease. *Current Opinion in Neurology*, 17(1):43.

Trottier, L. and Pratt, J. (2005). Visual processing of targets can reduce saccadic latencies. *Vision Research*, 45:1349–1354.

Tychsen, L. and Lisberger, S. (1986). Visual motion processing for the initiation of smooth-pursuit eye movements in humans. *Journal of Neurophysiology*, 56(4):953–968.

van Der Geest, J. N. and Frens, M. A. (2002). Recording eye movements with video-oculography and scleral search coils: a direct comparison of two methods. *Journal of Neuroscience Methods*, 114(2):185–95.

van Donkelaar, P. (1999). Spatiotemporal modulation of attention during smooth pursuit eye movements. *Neuroreport*, 10(12):2523–6.

van Donkelaar, P. and Drew, A. S. (2002). The allocation of attention during smooth pursuit eye movements. In Hyönä, J., Munoz, D. P., Heide, W., and Radach, R., editors, *Progress in Brain Research Volume 140. The Brain's eye: Neurobiological and clinical aspects of oculomotor research*. Elsevier Science, Amsterdam, Netherlands.

Van Gelder, P., Anderson, S., Herman, E., Lebedev, S., and Tsui, W. H. (1990). Saccades in pursuit eye tracking reflect motor attention processes. *Comprehensive Psychiatry*, 31(3):253–60.

Van Gelder, P., Lebedev, S., Liu, P. M., and Tsui, W. H. (1995a). Anticipatory saccades in smooth pursuit: task effects and pursuit vector after saccades. *Vision Research*, 35(5):667–78.

Van Gelder, P., Lebedev, S., and Tsui, W. H. (1995b). Predictive human pursuit and "orbital goal" of microstimulated smooth eye movements. *Journal of Neurophysiology*, 74(3):1358–61.

van Gisbergen, J. a., Robinson, D. a., and Gielen, S. (1981). A quantitative analysis of generation of saccadic eye movements by burst neurons. *Journal of Neurophysiology*, 45(3):417–42.

van Os, J. and Kapur, S. (2009). Schizophrenia. *Lancet*, 374(9690):635–45.

Verbruggen, F. and Logan, G. D. (2008). Response inhibition in the stop-signal paradigm. *Trends in Cognitive Sciences*, 12(11):418–24.

Vercher, J. and Gauthier, G. (1992). Oculo-manual coordination control: ocular and manual tracking of visual targets with delayed visual feedback of the hand motion. *Experimental Brain Research*, 90(3):599–609.

von Helmholtz, H. (1867). *Handbuch der physiologischen Optik*. Leopold Voss, Leipzig.

Wade, N. J. and Tatler, B. W. (2005). *The Moving Tablet of the Eye*. Oxford University Press, Oxford, UK.

Walker, R., Walker, D. G., Husain, M., and Kennard, C. (2000). Control of voluntary and reflexive saccades. *Experimental Brain Research*, 130(4):540–4.

Walton, M. M. G. and Gandhi, N. J. (2006). Behavioral evaluation of movement cancellation. *Journal of Neurophysiology*, 96(4):2011–24.

Wang, P. and Nikolić, D. (2011). An LCD Monitor with Sufficiently Precise Timing for Research in Vision. *Frontiers in Human Neuroscience*, 5(August):85.

Ward, D. J. and MacKay, D. J. C. (2002). Fast hands-free writing by gaze direction. *Nature*, 418(6900):838.

Watson, A. B. and Pelli, D. G. (1983). QUEST: A Bayesian adaptive psychometric method. *Perception & Psychophysics*, 33(2):113–120.

Wenban-Smith, M. G. and Findlay, J. M. (1991). Express saccades: is there a separate population in humans? *Experimental Brain Research*, 87:218–222.

Werner, W. (1993). Neurons in the primate superior colliculus are active before and during arm movements to visual targets. *The*

European Journal of Neuroscience, 5(4):335–40.

Wertheim, T. (1894). Über die indirekte Sehschärfe. *Zeitschrift für Psychologie und Physiologie der Sinnesorgane*, (7):172–187.

White, R. W. (1959). Motivation reconsidered: the concept of competence. *Psychological Review*, 66:297–333.

Wurtz, R. and Goldberg, M. E. (1972). Activity of Superior Colliculus in Behaving Monkey. III. Cells Discharging before Eye Movements. *Journal of Neurophysiology*, pages 575–586.

Wyatt, H. J. and Pola, J. (1987). Smooth eye movements with step-ramp stimuli: the influence of attention and stimulus extent. *Vision Research*, 27(9):1565–80.

Xia, R. and Barnes, G. (1999). Oculomanual coordination in tracking of pseudorandom target motion stimuli. *Journal of Motor Behavior*, 31(1):21–38.

Yanoff, M. and Duker, J. S. (2008). *Ophthalmology*. Mosby/Elsevier, Maryland Heights, MO, USA, 3rd edition.

Yarbus, A. L. (1967). *Eye Movements and Vision*. Plenum Press, New York, NY, USA.

Yeshurun, Y. and Carrasco, M. (1998). Attention improves or impairs visual performance by enhancing spatial resolution. *Nature*, 396(6706):72–5.

Yeshurun, Y. and Carrasco, M. (1999). Spatial attention improves performance in spatial resolution tasks. *Vision Research*, 39(2):293–306.

Zeki, S., Watson, J., and Lueck, C. (1991). A direct demonstration of functional specialization in human visual cortex. *The Journal of Neuroscience*, 17(March).

Zuber, B. L. and Stark, L. (1966). Saccadic Suppression: Elevation of Visual Threshold Associated with Saccadic Eye Movements. *Experimental Neurology*, 16:65–79.